Chemistry for Beginners

A. O. Hall

formerly Head of Science
Lanesborough School, Guildford

Heinemann Educational Books
London

Heinemann Educational Books Ltd
22 Bedford Square, London WC1B 3HH

LONDON EDINBURGH MELBOURNE AUCKLAND
HONG KONG SINGAPORE KUALA LUMPUR NEW DELHI
IBADAN NAIROBI JOHANNESBURG
KINGSTON EXETER (NH) PORT OF SPAIN

ISBN 0 435 64301 X
© A. O. Hall 1978
First published 1978
Reprinted with corrections 1979
Reprinted 1980

Printed and bound in Great Britain by
Morrison & Gibb Ltd, London and Edinburgh

Contents

Dedication

To my wife, Veronica, without whose love, patience, and constant encouragement this book would not have been achieved.

Acknowledgements

I wish to thank the various readers, particularly Mr M. J. Denial, whose detailed examination of both the original and expanded texts and advice were a great asset. I am also much indebted to Mr Martyn Berry for his helpful comments, and to Mr Hamish MacGibbon and his staff for their work in preparing the text for publication. Mr S. F. Swayne, Principal of Lanesborough Preparatory School, permitted the use of a preliminary edition of the book in the school, and I am most grateful to him for the useful experience gained. Finally, I want to thank my wife for her valuable criticism and painstaking typing and proof checking.

Acknowledgements for permission to publish photographs are due as follows: Barnaby's Picture Library, p. 66, p. 81, p. 114; BOC Ltd., p. 73 (Picadilly), p. 76 (oxygen), p. 99; British Airport Authority Training School, Stanstead, p. 76 (carbon dioxide), p. 97; British Airways, p. 6, p. 84; British Council, Press Department, p. 8; British Petroleum Company Ltd., p. 13, p. 44; British Tourist Authority, p. 40; British Steel Corporation, p. 33, p. 124; Central Electricity Generating Board, p. 108; Chemistry Society: *You and Chemistry*, p. 1 (laboratory); Department of Environment, p. 62; Greater London Council, Photographic Library, p. 31; Institute of Geological Sciences, p. 1 (moon rock); Keystone Press Agency, p. 94; Permutit Company Ltd., p. 7; Popperfoto, p. 70; Science Museum, p. 58; UKAEA, p. 103; Worcester Royal Porcelain Group, p. 21.

Preface

This is a pupils' book. It is based on the chemistry of Stage I of the Nuffield O-Level Sample Scheme. Some topics are, however, discussed in more detail, several additional ones have been introduced, and in one or two respects the order of presentation differs from the Nuffield proposals. The book constitutes a course of approximately three years and is suitable for secondary, middle, and preparatory schools.

An attempt has been made to preserve the essential pupil-discovery basis of practical work and at the same time to provide an elementary reference book. Each chapter therefore contains an investigational section comprising discussion material and detailed guidance on experiments, and a reference section in which the pupil can if necessary check his results and the knowledge to be gained from his investigations; there is also further information on each topic for the more advanced student, and with this in mind some of the questions have been made more demanding.

The book can, of course, be used in different ways, depending on the requirements of the teacher and his particular teaching methods. Those who do not normally use a textbook at this level may decide to use the book as a laboratory manual and basis for class discussion, the second part of each chapter only being used, for example, by selected pupils during study periods. Others, who prefer to encourage children in the guided use of textbooks, will wish them to have complete access to the book. It is hoped that the method of presentation and sequence of ideas will be helpful to those pupils who feel a need and desire for private study, especially at examination time.

Certain experiments are headed as demonstrations for the usual reasons of safety, technique required, or the nature of apparatus. Teachers will also wish to demonstrate one or two of the class experiments (such as Experiment 8.7) when pupils are not considered sufficiently experienced and reliable.

A chapter on equations (mainly for secondary schools) and one or two other topics in the book, are at present not required by preparatory schools, but the remainder of the text covers the

Common Entrance syllabus as described in the *Guide to Science Teaching in Preparatory Schools*.

Safety in the laboratory has necessarily become a subject of great emphasis, and a short appendix on this has therefore been included.

1978   A. O. H.

1 What is Chemistry?

When you are wondering what chemistry is all about you will probably think first of a chemist's shop, and decide that a chemist is someone who makes up prescriptions for medicines and sells things like pills, soap, and toothpaste. This kind of chemist, although he has had to qualify by passing chemistry examinations, is really a pharmacist. What other kinds of chemists are there? The analyst or analytical chemist is well known. He is consulted about things like food poisoning, or he may have to test water to find out whether it is safe to drink. Soils are sometimes analysed in case they contain chemicals which might be injurious to plants, and even rocks from the Moon were tested to see if they contained the same kinds of chemicals that we have on Earth. And there are research chemists who discover how to make entirely new kinds of substances like plastics, fibres, new drugs, special alloys for jet engines and spacecraft, etc.

There are many other types of chemists too, specializing in particular subjects such as petroleum products, paints, and horticultural sprays, but what have all these chemists in common? They are all concerned with chemicals—pure substances with special names like sulphuric acid, copper sulphate, chlorine. Obviously, they have to know how these pure substances (usually found in laboratories, chemical works, chemists' shops) are obtained from the ordinary substances, such as wood, coal, earth, and oil, which

A piece of moon rock

A modern chemistry laboratory

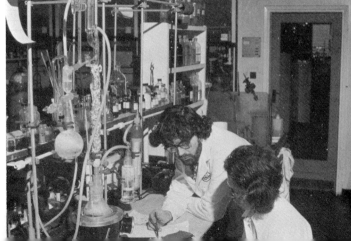

we see around us. They also need to know what happens when chemicals are mixed together. Will there be a big bang? Or perhaps an entirely new and useful substance will be made. This knowledge about pure substances is what you learn in chemistry.

The early chemists had to spend a great deal of their time finding out what ordinary substances contain. For instance, by heating coal, they discovered that it split up into many important chemicals such as benzene, ammonia, and carbon. They found that a number of valuable substances, such as salt, are dissolved in sea water. For many years it was thought that pure salt contained nothing else, just as pure aluminium contains nothing but aluminium. They had tried heating the salt until it melted, and dissolving it in water, but it did not change. Then it was discovered that if an electric current were passed through melted salt, the chemical split up into a soft, silvery metal (sodium) and a poisonous, green gas (chlorine). And it was soon found that the metal and gas could be made to form ordinary salt again. Other experiments were done with them. The metal floated on water; it fizzed and darted about, getting rapidly smaller until it disappeared completely. And the gas allowed a candle to burn in it, giving out a strange, reddish light. Eventually, chemists learned how to make new substances from the metal and gas in salt, such as anaesthetics, plastics, and a liquid for dissolving gold and silver.

In your study of chemistry you have to tread the same path as these early chemists, discovering things for yourself. You, too, will learn how to separate pure chemicals from mixtures of several substances, what happens when chemicals are heated or added together, how metals are obtained from their ores. The best way to learn chemistry is to do as many experiments as you can, and to find out *why* and *how* they work. Naturally, you will depend on your teacher for a full explanation of most things, and you should study your textbook carefully and constantly. Perhaps there are interesting books on chemistry in your school or local library. Some suitable ones are listed at the end of this book.

2 Chemical Magic

Children often become interested in chemistry through chemistry sets, crystal growing sets, and chemistry magic sets, even before they have started to learn chemistry at school. Perhaps you possess one or other of these sets. You have probably experimented with invisible ink, turned 'wine' into 'ink', or bounced bubbles on a heavy, invisible gas. The larger chemistry sets enable you to do quite a number of simple and colourful experiments, and although the chemicals provided are safe ones you should always carry out experiments very carefully because accidents can happen through broken glass, burns from the spirit flame, etc. And *never* indulge in the dangerous practice of inventing your own experiments. In any case, there are very good reasons why you should get into the habit of being careful and thoughtful, for in the school laboratory you will do experiments which can cause injury if they are not performed correctly and with proper precautions; also, success in science depends very much on your powers of *observation* and *deduction*. In simple words, when you do an experiment, it is essential to note everything that happens and then try to work out *why* it happens.

When you are a beginner, some of the chemical experiments you have to do at school are not always as colourful and exciting as you might wish. You have to learn about apparatus, laboratory technique, how to write up experiments, etc.

To whet your chemical appetite and kindle your interest, your teacher will probably show you some of the following experiments. They are called *demonstrations* because they are done by the teacher and not the class.

Demonstration 2.1. 'Wine into water'

Bubble sulphur dioxide through a pink solution of potassium permanganate. What happens? Since the quantity of gas required is small, it can be conveniently made by the action of dilute sulphuric acid on sodium sulphite.

Demonstration 2.2. The 'brown mushroom' experiment

In a fume cupboard, cover the bottom of a flask with sodium thiosulphate crystals. Add a few cubic centimetres of concentrated nitric acid. After the reaction is over, add water to the flask and empty quickly so that the large piece of sulphur is washed out. (Note: The short delay in this reaction is a useful discussion topic later—Chapter 10.)

Demonstration 2.3. The fountain experiment

The apparatus is shown in Figure 2.1. Blue litmus solution is placed in the lower flask and the *dry* upper flask is filled with hydrogen chloride and quickly placed in position. Your teacher may explain to you why the blue liquid rises up the glass tube and changes colour.

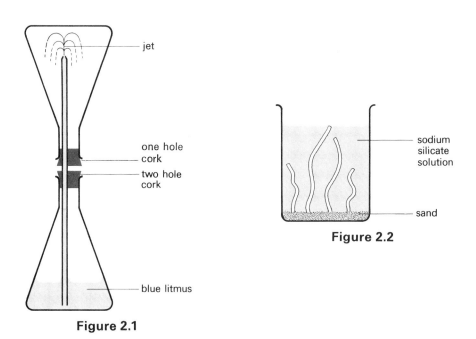

jet

one hole cork

two hole cork

blue litmus

sodium silicate solution

sand

Figure 2.2

Figure 2.1

Demonstration 2.4. A chemical garden

Pour a concentrated solution of sodium silicate (water glass) into a large beaker having a layer of sand on the bottom. When the sand has settled again, add single crystals of cobalt chloride, copper sulphate, iron sulphate, manganese sulphate, nickel sulphate, and magnesium sulphate. The crystals react with the solution to form silicates which 'grow' in rather an interesting way. Allow two or three days for the 'garden' to form completely.

The following experiments you can do yourself.

Experiment 2.5

Put about five cubic centimetres of ammonia solution in a test tube and add a few drops of litmus solution. Now add a little dilute acid until the colour changes.

Experiment 2.6

Place about five cubic centimetres of blue copper sulphate solution in a test tube, and then slowly add ammonia solution, one drop at a time. Note the effect, and then add more ammonia and shake the test tube until a clear liquid is obtained.

And here are two experiments that can be done at home.

Experiment 2.7

Coil a length of wire round a short piece of candle, leaving enough wire so that the candle, when lighted, can be lowered into a milk bottle. Now place about a dessert-spoon of baking soda and the same quantity of vinegar in the milk bottle. The 'fizzing' of the two substances produces a heavy invisible gas. Light the candle and lower it carefully into the bottle. What happens when it reaches the layer of heavy gas?

Experiment 2.8

Pour a little of the pink liquid from a jar of pickled red cabbage into a glass or cup. Add baking powder or washing soda until the pink liquid changes colour. Now add crystals of citric acid (used for making home-made lemonade) until the colour changes again. Do not taste the liquid.

Before you do any more experiments yourself, read carefully the Appendix on Safety Precautions (page 147). You should read it several times during your chemistry course and learn its contents.

3 Separating the Gold from the Sand

Discussion

Figure 3.1

The early gold diggers had to work out a way of separating tiny pieces of gold from sand. In the process known as 'panning for gold' the sand was swirled round and round with water in a pan until the lighter sand and other minerals were washed out, leaving the gold in the bottom. Much quicker methods of separation were eventually discovered.

Today, North Sea oil is often in the news. Crude oil contains many different substances, and a number of valuable products, such as petrol and engine oil, are separated from it.

An oil rig in the North Sea

All pure substances have to be refined, which means they have to be separated from mixtures of several substances. Think about any methods you know of separating things, and then discuss them. You may have found out something about the rather complicated equipment which keeps the water clean in a swimming pool. How are bits of paper, leaves, etc. separated from the water? The engines of cars and motor cycles have filters. Why? How can soil be separated from stones, twigs, etc.? Could you make a simple sieve for sifting things like soil and coal dust? Can you think of other things which may have to be sifted or filtered at home?

Sometimes a hard deposit called 'fur' forms inside a kettle. Have you ever noticed it and wondered how it gets there? There are chemicals in the water and these gradually build up as a deposit when the water is boiled and turned into steam. Can you think of a way to obtain salt from sea water? Sugar and salt both dissolve in water. Does sand dissolve? How could you get pure sand from salty sand? Rock salt consists of large crystals of dirty salt which have come from a salt mine or the sea. Try to work out a method of refining rock salt in order to obtain pure, dry salt. When ice forms on top of sea water in arctic regions, do you think the ice is pure or mixed with salt?

You probably know that car batteries need distilled water. Distilled water is very pure water. In what way do you think it differs from tap water, and why cannot tap water be used in batteries? Steam comes from water, so it must be the gas of water. How could you turn the steam back into a liquid and collect the water again? Do you think that steam from dirty water would be dirty? You may have visited a petroleum oil refinery or read about one. Did you manage to find out the main process used for getting petrol, diesel fuel, and other products from the crude oil?

'Fur' which has
formed in a
hot-water pipe

Instruction

Mixtures

Most of the ordinary things we see around us are *mixtures* of several different substances. Sea water has many valuable substances in it as well as salt. Even river water contains dissolved chemicals. Clean air is a mixture of nine different gases, and in towns other gases are added to it from chimneys, cars, and factories.

Polluted air above Stoke-on-Trent, the centre of the pottery trade

Earth, coal, oil, wood, and cloth are all mixtures, like air. The object of the various separation methods you are now going to try out is to show you how pure substances can be obtained from mixtures; for example, how salt can be got from sea water. When substances are separated like this from mixtures, they are said to be purefied or refined. In the chemical industry the methods are naturally of first importance, and at school they have to be thoroughly understood to enable you to carry out other experiments.

Methods of Separation

Filtration

A mixture of a liquid and a solid not dissolved in it is separated by filtration. Such a mixture might be water and mud, for example. Filter paper, which is like blotting paper, is used. It allows a pure liquid like water to pass through it but traps any solid particles like those of mud or sand. A round *filter paper* is folded in half so that it is the shape of a semi-circle and is then folded in half again so that it is a quarter of a circle (a quadrant). It can then be carefully opened out into a cone which fits into a *filter funnel*. Before it is placed in the funnel the latter is rinsed with water. This not only cleans the funnel but also makes the paper stick to it. The liquid to be filtered is poured slowly down a glass rod held so that one end is inside the funnel. This prevents any splashing and ensures that the liquid does not get between the paper and the funnel. Be very careful not to touch the filter paper with the glass rod or with your fingers because it is easy to make a hole in it when it is damp. The solid (e.g. mud) remaining on the paper is called the *residue*, and the clear liquid which passes through the filter paper is called the *filtrate*. The residue should be washed by pouring hot water through it, and it can then be dried by opening out the filter paper carefully and placing it on a wire gauze over a *very small* Bunsen flame.

Experiment 3.1. To separate a mixture of water and mud

First, try the quick way of letting mud settle in the bottom of the test tube and then *decant* the clear water on top by simply pouring it into another test tube or beaker. You will not be able to obtain all the water this way. Now use the filtration method, being careful to pour the muddy water gently down the glass rod into the funnel without touching the filter paper. This time you can collect all the clean water. Figure 3.2 will help you in assembling the apparatus.

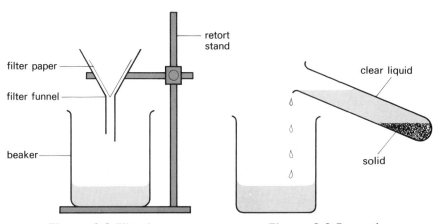

filter paper

filter funnel

beaker

retort stand

clear liquid

solid

Figure 3.2 Filtration apparatus   **Figure 3.3** Decanting

Evaporation

If a mixture of a liquid (e.g. water) and a solid dissolved in it (e.g. sodium chloride (salt)) has to be separated, *evaporation* is the process used, provided only the solid is required. This simply consists of boiling the mixture in an evaporating basin so that all the water is driven off as steam, leaving the pure solid as a residue in the basin (see Figure 3.4). To prevent the solid from spurting during the final stages of evaporation, heating is continued over a steam bath (a beaker of boiling water). If pure water is required from the mixture (the salt water), the process of *simple distillation* is employed (page 11).

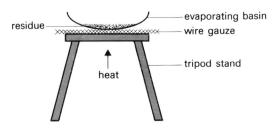

residue — evaporating basin — wire gauze — tripod stand — heat

Figure 3.4 Evaporation

Experiment 3.2. To separate a mixture of copper sulphate and sand

Copper sulphate dissolves in water, but sand does not. Put the mixture in a small beaker and add some water. Heat gently on a wire gauze and tripod stand over a Bunsen burner. Be very careful when using a Bunsen burner. You will learn more about it when you have been taught more chemistry (Chapter 11). Close the air hole before lighting the burner and then open it a little so that there is only a little tip of yellow in the flame. Stir with a glass rod to help the copper sulphate to dissolve, and when a little steam appears but the beaker is not too hot to hold, filter into an evaporating basin. Wash the residue of sand on the filter paper with a little warm water, and then dry it as described on page 9.

Evaporate the filtrate until its colour becomes deep blue and crystals form when a drop of the liquid is placed on a glass rod. Then pour some of the liquid into a shallow dish in order that crystals of copper sulphate can slowly form.

Experiment 3.3. To purify rock salt (impure sodium chloride)
Place the crushed crystals of rock salt in a beaker and add
about half a beaker of water. Warm and stir until the salt has
dissolved. Filter into an evaporating basin and evaporate the
filtrate (salt water) until about half the liquid remains, then
continue evaporation over a steam bath (Figure 3.5).

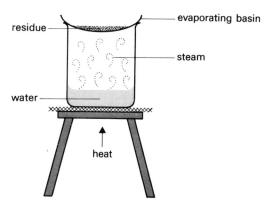

residue — evaporating basin

— steam

water —

heat

Figure 3.5 Evaporation on a steam bath

Simple distillation

When water contains a substance, such as sodium chloride (salt),
dissolved in it and it is necessary to obtain pure water from this
mixture, the process of *simple distillation* is employed. The liquid is
boiled so that the water becomes steam, which is then condensed
(liquefied by cooling) and collected in another container. The
dissolved substance remains, with a little water, in the distillation
flask. Water can be obtained from ink in this way.

**Experiment 3.4. To obtain pure water from a mixture of
copper sulphate and water using simple distillation
apparatus**
You require a small conical flask fitted with a side arm and a
cork, a bent glass tube, a short piece of rubber for connecting
flask and tube, a test tube, a beaker, and the usual heating
apparatus. Assemble it as in the diagram (Figure 3.6), then
place some of the mixture in the flask, and replace the cork.
Nearly fill a beaker with cold water, put in a clean test tube,
and connect the tube from the flask to the test tube. Now
boil the liquid and quite soon the steam will condense in the
test tube as pure, distilled water.

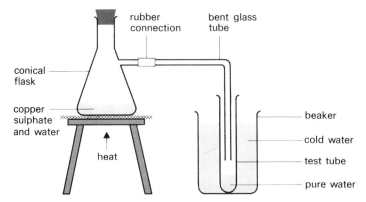

Figure 3.6 A simple apparatus for distillation

Demonstration 3.5

Your teacher will show you how a proper distillation apparatus works. In this, a Liebig condenser is used to condense the steam more thoroughly. Note that the water from the tap circulates through the condenser and goes into it at the lower end. Why? (See Figure 3.7.)

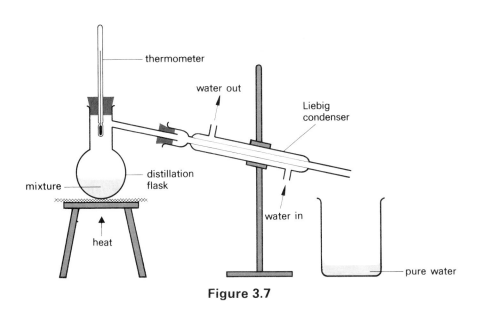

Figure 3.7

Fractional distillation

Although simple distillation can be used for separating a mixture of two liquids, the method is not very efficient unless the boiling points of the liquids are very different. For separating liquids fractional distillation is employed. The process consists of vaporizing the parts or fractions of the mixture and condensing and collecting each one separately. A simple example is a mixture of ethanol and water. Ethanol boils at only 78 °C, whereas the boiling point of water is 100 °C. So when the mixture is distilled almost pure ethanol is collected first, when the thermometer reads 78 °C; then, after more heating, ethanol with a little water is collected, followed by water with a little ethanol; finally pure water boils off at 100 °C. The ethanol/water fractions can be redistilled. Petrol, paraffin, lubricating oils, and other petroleum products are separated from crude petroleum by fractional distillation. This is done in a gigantic distillation apparatus which includes a tall tower about thirty metres high called a fractionating column (see Chapter 13). Gases in the air are separated by fractionally distilling liquid air (see Chapter 8). The laboratory apparatus for fractional distillation is the same as that shown in Figure 3.7 except that the distillation flask is fitted with a small fractionating column (Figure 3.8), consisting of a glass cylinder filled with glass beads or small pieces of glass. This device helps to separate the different vapours in the following way. Let us take the example of a mixture of equal volumes of water and ethanol. When the mixture is boiled, the more volatile ethanol vaporizes fairly easily but a little water vapour is also formed. So there is a mixture of vapours in the flask, say 70% ethanol and 30% water. When this mixture rises up the fractionating column it condenses on the cooler beads and begins to flow back into the flask. But it meets hotter vapours rising from the heated liquid which is steadily increasing in temperature. This causes the

A fractionating column at BP's oil refinery at Rotterdam

condensed mixture on the beads to vaporize again, and the resulting vapour again contains a higher proportion of ethanol than in the liquid mixture, e.g. 85% ethanol. This process of condensation and vaporization continues right up the fractionating column (which is hot at the bottom and cooler at the top), the proportion of ethanol becoming progressively higher until eventually the vapour leaving the top of the column is pure ethanol.

***Demonstration 3.6*. To separate a mixture of ethanol (industrial methylated spirit) and water by fractional distillation**

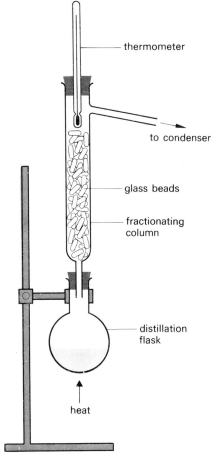

Figure 3.8 A fractionating column

Make a mixture consisting of about three parts of water to one of ethanol, and add some to the flask so that it is not more than about a third full. Boil very gently to avoid the fractionating column becoming flooded. While the first fraction is being collected, show that the mixture will not burn by adding a few drops to an evaporating basin and trying to ignite the liquid. Collect fractions boiling at about 78 °C, 82 °C, 90 °C, and 100 °C, in separate beakers, and test each to see if they will burn.

Fractional crystallization

None of the methods so far described can be used to separate two or more solids mixed together when they all dissolve in water. In this case, *fractional crystallization* is the process to be adopted. You have probably made crystals in physics, and you will understand that when a concentrated (strong) solution of a substance (e.g. copper sulphate) is cooled and allowed to evaporate slowly crystals of the substance are formed. If the solution contains say, two substances, the crystals of one usually form before those of the other. You will understand this better when you have studied Chapter 6. So the mixture is simply dissolved in hot water and then the liquid is slowly cooled in a shallow dish. When the crystals of one substance form, the remaining liquid is poured off (decanted) into another dish and left for the other crystals to form. The first crystals are washed with distilled water, carefully and quickly, so that they do not dissolve. It is sometimes necessary to heat the solution containing the second substance in order to concentrate it by evaporation.

Experiment 3.7. **Separation of a mixture of copper sulphate and potassium chlorate by fractional crystallization**

Make a little of the mixture by adding together about 2 g of each compound, and place it in about 50 cm³ water in a beaker. Stir until the mixture has dissolved except for a little on the bottom. Now warm gently until this too has just dissolved. Transfer some of the liquid into a shallow dish and leave it to cool. After a few hours white crystals of potassium chlorate will be formed. Pour off the blue liquid into another dish for the copper sulphate crystals to form, and wash the white crystals very carefully with a very small quantity of distilled water. Place them on a filter paper to dry. The blue copper sulphate crystals may take a day or more to form.

Sublimation

One of the two solids may be a substance which *sublimes*, i.e. passes from solid to gas without becoming a liquid. If this is so, the two can be separated by heating the mixture in a suitable container so that the one which sublimes turns into a gas. A cool glass surface above the container causes the gas to condense directly back to a solid again. This process of separation is called *sublimation*.

Experiment 3.8. To separate a mixture of two solids, which both dissolve in water, by sublimation

Mix together about 2 g each of sodium chloride (salt) and ammonium chloride and put the mixture in a test tube. Using a test tube holder heat the mixture gently in the Bunsen flame, moving the test tube all the time so that it will not crack. Ammonium chloride is a substance which sublimes, so after a few minutes you will see a white collar of it on the cooler, upper part of the test tube. It is obviously important that you heat only the lower part of the test tube. Although the two substances cannot be collected separately in this small experiment, you are able to see how separation occurs.

Chromatography

Many substances, particularly those derived from living things (proteins are an example), are so similar in certain properties (characteristics) such as boiling point, that they cannot be separated from a mixture by any of the above processes. The purification can then often be done by *chromatography*. It is a very simple process based on the fact that different substances creep along filter paper or other porous material at different rates, even when the substances are very similar in other respects. Thus, if a drop of a solution of two or more substances is placed in the middle of a filter paper, the substances will separate as the solution creeps outwards towards the edges. If they happen to be coloured substances you will soon see rings of different colours. When the method was first discovered, coloured substances were being tested, so it was called chromatography, a name derived from two Greek words meaning 'colour writing'.

Experiment 3.9

Mix together about 10 cm³ of blue litmus solution and 5 cm³ of methyl orange solution in a small beaker. Place a filter paper on a dish or beaker. Now put one drop of the mixture on the centre of the filter paper, using a teat pipette. When the drop has finished spreading, add another, and repeat until three or four drops have been added. Do the liquids separate? Methyl orange and litmus are solid substances which, for laboratory use, are dissolved in a *solvent* (Chapter 6). As the solvent (water in this case) moves through the porous filter paper, the substances dissolved in it move at different rates. Thus, the methyl orange gets 'left behind' because the litmus moves faster.

Experiment 3.10

Repeat Experiment 3.9, but use a mixture of liquids called Universal Indicator.

Experiment 3.11. To separate chlorophyll and xanthophyll from the green colouring matter in grass

For this experiment you need a pestle and mortar, an apparatus which is used mainly for crushing lumps of a solid substance into a powder. Place a handful of grass clippings in the mortar with a little water and then try to make the green colouring matter dissolve in the water by grinding the grass with the pestle. Soon you will obtain a green liquid, but has the green colouring matter really *dissolved* in the water? To see if it has, filter the liquid. What happens? Do you think water is a good solvent for this experiment? Now try another solvent, called propanone (acetone). Add about half a test tube of it to some more grass in the mortar, and grind the grass again until you have made a dark green liquid. Filter a few drops of this liquid to see if the green colouring has really dissolved. If you are satisfied that it has, pour the green liquid into a small beaker. Make two cuts, with scissors, from the edges of a filter paper to the centre, so that a strip of paper about half a centimetre wide is produced. Bend this strip downwards at right angles to the rest of the paper, and then place the filter paper on a small beaker containing propanone, with the strip dipping into the liquid (Figure 3.9). Now take a teat pipette and add four drops

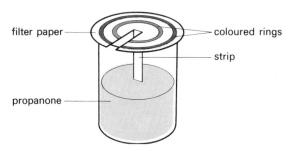

Figure 3.9

of the green liquid to the centre of the filter paper as you did in Experiments 3.9 and 3.10. The propanone, creeping up the 'tongue' of paper, helps the green liquid to spread outwards and separate into the two substances. Your filter paper, with the two substances separated as coloured rings, is called a *chromatogram*. If the coloured rings are cut out from several chromatograms they can be placed in two beakers, one for each colour. A little warm propanone is then added to each beaker and the mixture is stirred. A yellow liquid containing the xanthophyll and a green one of chlorophyll are obtained. Note that *propanone* must be *handled with care* as it is *highly flammable*.

Separation of gases

Gases often have to be separated from a mixture of gases, a well known example being the separation of oxygen from air (page 13). There is a simple, quick way of removing a gas from a mixture of gases if the gas to be removed dissolves easily in some liquid. The mixture is bubbled through the liquid and the gas which dissolves remains in the liquid. The air contains small quantities of the gas, carbon dioxide, and this can be separated by bubbling air through a liquid called sodium hydroxide solution, which absorbs the carbon dioxide.

Demonstration 3.12. **To separate a mixture of oxygen and nitrogen dioxide**

Heat lead nitrate crystals in a test tube fitted with a delivery tube leading to an inverted gas jar of water in a bowl (Figure

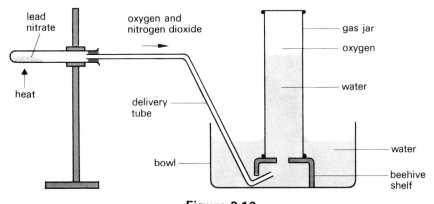

Figure 3.10

3.10). Can you see the two gases? Which one is oxygen?
Which one will relight a glowing splint of wood? Does one
dissolve in the water?

You now know how to obtain chemicals, such as sodium chloride
(salt), copper sulphate, and water, from mixtures. The next step is
to learn more about these pure chemicals. Can any of them be
separated themselves into other substances? The idea of a pure
substance containing others, somehow hidden within it, may seem
strange at first. Yet the early chemists always suspected that this
could be the case, and spent much of their time trying to extract
gold from many different chemicals. They realized that ordinary
methods of separating chemicals from mixtures would not help, so
they tried more drastic experiments such as heating a substance
strongly or mixing it with another chemical and then heating the
two together. In the next chapter you will investigate the effect of
heating substances.

Summary

Methods of Separation

1. Filtration
To separate a mixture of a liquid and an insoluble solid.

2. Evaporation
To separate a mixture of a liquid and a soluble solid when the
latter is required.

3. Simple distillation
To separate a mixture of a liquid and a soluble solid when the
liquid is required.

4. Fractional distillation
To separate a mixture of two or more liquids with different
boiling points.

5. Fractional crystallization
To separate a mixture of two or more soluble solids.

6. Sublimation

To separate a mixture of two soluble solids when one sublimes (changes, when heated, directly into a gas which condenses directly into a solid again when cooled).

7. Chromatography

A method of separating a mixture of substances possessing very similar properties and thus being difficult to separate by methods 1–6 above. The process depends on the fact that different substances move through filter paper or other porous materials at different rates when aided by a suitable solvent.

8. Separation of gases

A simple method is to bubble the mixture of gases through a liquid which dissolves or reacts with only one of them, allowing the other gas to pass on and be collected.

Questions

1. List all the items of apparatus needed in a filtration experiment.
2. Explain the terms *residue* and *filtrate* in filtration.
3. Why is a glass rcd used when filtering?
4. Describe how you would separate pure sodium chloride (salt) from salt water.
5. You are given a mixture of sand and copper sulphate. Describe how you would obtain dry sand and pure copper sulphate from it, explaining each step carefully.
6. How could you obtain pure water from sea water?
7. Outline the methods by which you could obtain samples of both sodium chloride (salt) and water from a mixture of both.
8. Describe briefly the process of fractional distillation, giving an example.
9. How can a mixture of two soluble solids be separated?
10. Describe briefly an experiment to separate two substances from the green colouring matter in grass. Name the substances and explain the function of the solvent used.
11. Having separated two substances from the green colouring matter in grass, how would you prepare solutions of each of them?
12. How can carbon dioxide be removed from air?

4 Hotting them up

Discussion and Investigation

Figure 4.1

What happens when substances are heated? There are many examples of things being heated at home; water in an electric iron to give steam, raw food to provide cooked meals. Evidently heat changes substances, a great deal in some cases. What other everyday examples can you think of? In potteries, mugs, vases, jugs, etc., made of soft clay, are baked until they are hard. Coal is heated to make coke and many other valuable products, and limestone is heated and turned into lime for gardens.

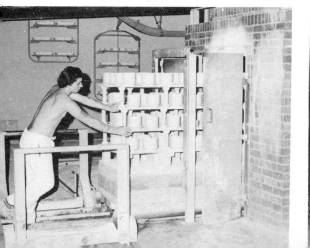

A trolley of plates being pushed into a kiln at 950 °C

A trolley of crockery emerging from a kiln after firing

One of the first things a chemist does when he wants to learn about or identify a substance is to heat it and observe any changes that may occur. He will heat a very small quantity first, with great care, in case the substance explodes, or behaves in some other dangerous way. The substance may alter in colour, for example, or ignite (catch fire). It may only change its state (turn into a liquid or gas).

If you possess a chemistry set you have probably already tried heating various chemicals. What happened when you heated blue copper sulphate? Discuss the effect of heat on the chemicals you have handled.

You are now going to investigate the effect of heating a few chemicals. When you carry out chemical experiments always remember:

(a) That many chemicals are poisonous or injurious to the skin. Some give off fumes, so you should never smell a chemical unless instructed to do so by your teacher. Sometimes you will be told to wear goggles during an experiment to prevent anything from spurting into your eyes, should you heat a substance too much or be careless in some other way.

(b) That chemicals have been carefully purified, and may not 'work' properly if they are contaminated, i.e. accidentally mixed with others. Therefore the same spatula should never be used for different chemicals unless it is first washed and dried. Stoppers of reagent jars and bottles (used for chemicals) must never be mixed up. The only way to avoid this is to replace the stopper as soon as you have taken what you need of the chemical.

(c) That chemicals are expensive, and only the smallest quantities possible should be used.

In the following experiments, you should record in your rough book everything that happens when the substances are heated. Careful and thorough observation is essential in science, and helps you to learn about the properties (characteristics) of substances. In particular, write down whether the substance has

(a) changed completely;

(b) changed only temporarily, i.e. returned to its original appearance when allowed to cool;

(c) not changed at all.

Experiment 4.1

Heat a little of each of the following substances separately in a crucible on a pipe-clay triangle over a *small* Bunsen flame:

copper sulphate (stir gently with a glass rod) ; sodium chloride (salt) ; zinc oxide (adjust burner to a fairly hot flame after initial heating) ; sugar. After a few minutes, remove the Bunsen burner and let the substance cool.

Experiment 4.2
Repeat Experiment 4.1 in exactly the same way, but with the following substances :
cobalt chloride (stir gently with a glass rod) ; anhydrous sodium carbonate ; copper carbonate ; powdered roll sulphur (heat only gently ; if it catches fire put on the crucible lid and stop heating). (Note : Naphthalene may be substituted for sulphur, observing the same precautions.)

Experiment 4.3
Heat a small piece of the following metals :
magnesium ribbon, copper foil, lead foil (in a fume cupboard). Hold the metal in tongs over a wire gauze or other non-flammable material. The Bunsen burner should be held at an angle so that nothing falls into it and the flame touches one side of the metal. Do not use a yellow flame, for this would deposit soot on the metal. And do *not* look too intently at the bright light of the burning magnesium. This can cause eye strain. Record your results as in the previous experiments.

Now discuss your experiments. Which substances did not change when heated? Which changed only while being heated, and then, when cool, were restored to their original appearance? With regard to the substances that changed completely, can you offer any explanation? Do you think any of them lost something, such as a gas, when they were heated? Or perhaps you may think that one or two gained something.

In the next two experiments you will investigate more thoroughly the effect of heating two of the substances tested in the previous experiments.

Experiment 4.4. To find out whether blue copper sulphate gains or loses mass when heated
Half fill a test tube with the copper sulphate powder and then weigh it carefully on a lever arm balance. Note the mass (in grams) in your rough book. Now heat the copper sulphate gently, using a test tube holder, *and moving the test tube* in the flame so that the glass does not crack. When all the blue colour has gone, stop heating and allow the test tube to cool by standing it up in a *clean* dry beaker. Then weigh the tube and contents again. Did the copper sulphate gain or lose mass ?

Experiment 4.5. To find out whether magnesium gains or loses mass when heated

Place about half a gram (roughly 30 cm) of magnesium ribbon (coiled up) in a crucible, replace the lid, and weigh. Record the mass of crucible, lid, and magnesium in your rough book. Put the crucible on a pipe-clay triangle and heat it gently on a tripod stand. Remove the lid occasionally to allow air to enter the crucible. When the magnesium ignites, continue to lift the lid every now and then, but only a little way so that few of the white fumes escape. When the burning is over, heat the crucible strongly for about two minutes with the lid off. Be careful not to let any of the white powder drop from the lid. Replace the lid and allow the crucible to cool. Then weigh the crucible, lid, and contents again. Has the magnesium gained or lost mass?

Let us discuss the last two experiments. What was the colour change when you heated the copper sulphate? Did you notice anything else? Why do you think the compound lost mass? In Experiment 4.6 you can explore this question further. How would you describe what happened when you heated the magnesium? Suppose it had been heated with the lid on all the time, would the result have been the same? How do you explain the change in mass?

Experiment 4.6. To investigate what is lost when copper sulphate is heated

Half fill a test tube with copper sulphate, fit a cork and delivery tube, and clamp the test tube at a slight angle in a retort stand.

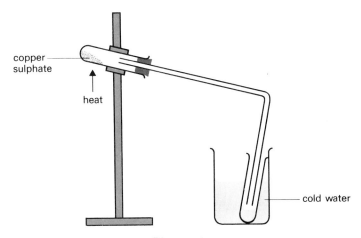

copper sulphate

heat

cold water

Figure 4.2

The end of the delivery tube dips about half way into another test tube sitting in a beaker of cold water (see Figure 4.2). Heat the copper sulphate carefully as in Experiment 4.4 until it has gone white and you have collected some colourless liquid in the second test tube. Remove the Bunsen burner, but leave it alight. Now investigate the liquid in the test tube. Clamp the test tube vertically in a retort stand with a thermometer clamped above it. The bulb of the thermometer should be just above the liquid in the test tube. Heat the liquid very gently until it boils. What is the temperature shown on the thermometer? Take a little of the *white* powder from the first test tube and place it on a watch glass held in the palm of your hand. Now add a few drops of the liquid to the watch glass. What happens to the white powder? Did the watch glass get warm? Examine a drop of the liquid on your finger. What do you think the liquid is?

Add some of the white powder to another colourless pure liquid such as propanone (acetone).

The magnesium turned white when heated and gained in mass. The copper became coated with a black powder (Experiment 4.3). Did they take something from the air?

Experiment 4.7. To find out whether copper turns black when heated in the absence of air

Make a sealed 'envelope' from a piece of copper foil by folding it carefully and flattening the edges together with a hammer so that air cannot get inside. It helps if the foil is cut into a convenient shape with a pair of scissors:

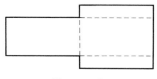

Figure 4.3

Heat the envelope in a Bunsen flame until it turns black on the outside. Allow it to cool and open it. Is the inside black? Was the air necessary for the copper to form the black powder? Do you think that the copper and black powder weigh more than the original piece of copper before heating?

Explanation and Further Information

When you heat a substance in the laboratory one of several things happens:

1. It changes state.
2. It temporarily changes in appearance, such as colour, but does not change in any other way.
3. It splits up (decomposes) into other substances which are quite different.
4. It joins up (combines) with another substance to form a different substance.
5. It does not change in any way.

A *chemical reaction* is said to occur when a chemical changes into a new substance, as in (3) and (4), and the result of the chemical reaction is a *chemical change*. You will learn more about chemical reactions in Chapter 10.

Let us now consider the results of all your experiments.

1. When *sulphur* is heated, it changes state, becoming a liquid which has different colours and is thick or runny, depending on how hot it is. When the liquid sulphur cools it changes back into solid sulphur again. *Lead* also melts and the liquid soon becomes solid lead again when cooled.
2. The colours of sulphur when heated are only temporary. *Zinc oxide* behaves in a similar way, going yellow when hot.
3. The substances which decomposed were *blue copper sulphate* which turned white; *sugar*, which became brown or black; *cobalt chloride* went blue, and *copper carbonate* went from green to black. All these colour changes remained when the substances cooled. There are, of course, many substances which decompose without showing colour changes, and other ways of detecting decomposition.

   In Experiment 4.6 you found that blue copper sulphate split up into a white substance and a colourless liquid which was water. Have you any suggestions as to how any of the other substances split up?
4. The *magnesium* gained in mass and absorbed air or something in the air, forming a white powder. You also found that copper needed air when it formed a black powder. Most metals behave like this when heated in air; for example sodium forms a yellow powder and mercury a red one. When you heated the *lead* it melted, but with more heating it would have formed a yellow powder.

5. The substances in your experiments which did not change at
   all were *sodium chloride* (salt), and *anhydrous sodium carbonate*.
In Experiments 4.4 and 4.6 you discovered that when blue copper
sulphate is heated it forms a white powder (called anhydrous copper
sulphate) and water, and that if water is added to the white powder
it turns back into blue copper sulphate, becoming quite hot in the
process. All this can be written more easily by using a *word equation*:

blue copper sulphate + heat ⇌ white copper sulphate + water

The arrows point in both directions to show that the things on the
left produce the things on the right, or vice versa. So when you
added water to the white copper sulphate, heat was given out as well
as blue copper sulphate being formed. This reaction is one of the
chemical tests for identifying the presence of water. But note carefully that any liquid containing water (for example, a dilute acid—
see Chapter 7) gives the same result. To prove that a liquid is
pure water further tests must be carried out (see below). Substances
do not always behave like this when they are heated, splitting up
into simpler substances which can react together to form the
original substances again. For example, the word equation to show
the burning of magnesium has only a single arrow:

magnesium + a gas → white powder + heat

This is because it is not possible to heat (add heat to) the white
powder to produce magnesium and a gas.
   The word 'anhydrous' is used in the names of some of the
substances with which you experimented. It is derived from two
Greek words and means 'without water'. Thus blue copper sulphate loses water when heated and becomes white anhydrous
copper sulphate. Similarly, pink cobalt chloride loses water to
become anhydrous cobalt chloride which is blue. You heated some
anhydrous sodium carbonate in Experiment 4.2. This substance is
formed by heating sodium carbonate (washing soda) so that it loses
all its water. When you hold crystals of a substance like blue copper
sulphate in your hand they do not feel damp. This is because the
water they contain is chemically combined in them. This kind of
compound is called hydrated, e.g. hydrated copper sulphate.
   A pure substance differs from a mixture because it has properties
which never vary. For example, when a substance is heated and
changes state it does so at a particular temperature if it is pure. The
purity of a chemical is often checked in this way. Pure water, for
instance, boils (turns into a gas) at exactly 100 °C at standard
pressure; if the boiling point of water is not 100 °C at standard
pressure the water must contain some impurity. Pure ethanol has a

boiling point of 78 °C. If ethanol contains any water, which is often the case, the boiling point is higher than this. Remember that the temperature of a liquid (heated in an open container) cannot go higher than its boiling point. After this, all the heat energy put into a liquid is used to change its state into a gas.

You have gained some useful knowledge about several important substances, chemical apparatus, and laboratory technique by doing the experiments in this chapter. Having learned how to separate pure substances from mixtures of things, you have discovered that some of these substances can split up into quite different ones, or can add things to themselves, like the magnesium and copper, forming powders of various colours. It is now time for you to find out how it is that substances can behave in this way.

Questions

1. Describe what you see and the type of change when the following substances are heated :
   (a) sulphur, (b) zinc oxide, (c) blue copper sulphate, (d) magnesium, (e) anhydrous sodium carbonate.
2. Repeat the instructions in Question 1 for :
   (a) lead, (b) sugar, (c) copper carbonate, (d) cobalt chloride, (e) sodium chloride.
3. Does magnesium gain or lose mass when heated ? Explain your answer.
4. When a sealed 'envelope' of copper foil is heated why does the outside but not the inside go black ?
5. What was the liquid you collected in a test tube by heating blue copper sulphate, and how did you identify it ?
6. Write a word equation to show what happened in Question 5, i.e. to illustrate the chemical reactions taking place.
7. Explain how a pure substance differs from a mixture.
8. One of several different things happens when a substance is heated. Describe three of these, giving examples.
9. How do pink cobalt chloride and blue anhydrous cobalt chloride differ ? What does 'anhydrous' mean ?
10. Explain the words *decomposes* and *combines*, giving examples to illustrate your answer.
11. Some green crystals of the salt iron sulphate were heated and turned white. Drops of colourless liquid collected near the mouth of the test tube. Can you name the white compound and the liquid ?
12. When the metal strontium is heated in a crucible and a white powder is obtained, does the powder weigh more or less than the metal before heating ? What is the name of the powder ? Explain how the powder was produced by heating the metal.

13. State which of the following changes involve a chemical reaction :
   (a) A silver metal wire turns red when held in a Bunsen flame and when cooled becomes silvery again.
   (b) Sodium chloride is dissolved in water. The solution of salt water is then evaporated and the salt recovered.
   (c) The metal sodium burns with a yellow flame and a pale yellow powder is formed.
   (d) Water is frozen into ice which is then melted into a liquid.
   (e) A white powder is heated strongly and turns a reddish brown colour.

5 Nature's Construction Set

Discussion

In your experiments to investigate the effect of heating substances you found that some of them changed completely and that new or different substances were produced. Which chemicals behaved in this way? Describe some of the new substances formed. Let us write word equations to show what happened when two of the substances were heated:

cobalt chloride + heat → anhydrous cobalt chloride + water
   (pink)   (blue)
copper carbonate + heat → copper oxide + carbon dioxide
   (green)   (black)   (gas)

You probably noticed the drops of water on the cooler parts of the glass rod when the cobalt chloride was heated. Carbon dioxide is a colourless gas, like air, so you would not have been aware of it.

Evidently some chemicals, although they are absolutely pure (not *mixed* with others), contain simpler chemicals which can be released by heating. Do you think that these other chemicals are also pure substances, having their own particular properties? Their names suggest that even these simpler chemicals contain others. You probably know already that water contains the gases hydrogen and oxygen. What is your opinion about copper oxide and carbon dioxide? Can this splitting up process continue? Can hydrogen and oxygen be split up into other chemicals?

The answer to the last question is a firm *no*. There are certain substances which cannot be decomposed (split up) any further, and two of them are hydrogen and oxygen. These simple substances make up all others and they are called *elements*. Can you name the elements in copper oxide? Carbon is an element. What are the elements in carbon dioxide? Now you can probably name the elements in copper carbonate.

Because elements are the basic ingredients of more complicated substances they are sometimes called the 'bricks of matter'. The pure chemicals which contain them are called *compounds*, examples being copper sulphate, cobalt chloride, carbon dioxide, and water.

This building up of compounds from elements is rather like the construction of a house. First, there are the basic materials such as bricks, wood, and iron. (These are like the elements.) They are needed for making the various main parts of the house: walls, water tanks, floors, etc. These main parts can be likened to the simpler compounds, such as copper oxide and carbon dioxide. Finally, all the parts are put together to make the house itself, just as simpler compounds can combine chemically to form more complicated ones, like copper carbonate. You could use a model construction set, such as Lego, to explain elements and compounds, or perhaps you can think of a different kind of analogy altogether. But be careful about *mixtures*. Although many mixtures, such as coal, wood, glass, and cement appear to be definite substances and to possess their own invariable properties, in fact they vary considerably in composition, i.e. in the elements and compounds they contain and in the proportions of these substances. So a mixture, having no set pattern, should be compared to miscellaneous piles of building materials rather than to a house.

Some of the many different substances used when a house is built

Explanation and Further Information

Elements

An element is a pure substance which cannot be decomposed into any other substances by chemical means.

The tiny particles, atoms, which make up an element are all the same kind. For example, the element iron contains only iron atoms. In Chapter 1 it was mentioned that ordinary salt (sodium chloride, a compound) contains two simple substances, the elements sodium (a metal) and chlorine (a gas, non-metal). So salt contains different kinds of atoms, sodium atoms and chlorine atoms. Other examples of elements are aluminium, lead, gold, and other pure metals. Carbon, sulphur, oxygen, and hydrogen are some of the non-metal elements. Some of these are solids, like sulphur, and others are gases, like oxygen. You will find a list of common elements at the end of this chapter.

Just as there are only a few basic or simple materials, such as wood, glass, and steel, in a building, so there are only about ninety elements which make up the world. Every substance, living or non-living, is composed of one or more of these elements. The gases in the atmosphere, the molten rocks in the centre of the earth, the plants and animals are all made from these ninety elements. Most of them are very rare, many being rarer than gold. Nearly all common substances are made of a mere twenty or so different elements. There are only *four* in the following compounds:

washing soda	water	sugar
glycerine	ethanol	starch
soap	citric acid	baking soda
carbon dioxide	cellulose	hydrogen peroxide

Some elements are found in the free state, i.e. by themselves. Examples are gold, carbon, sulphur, and various gases. But most are combined chemically together in compounds.

Compounds

A compound is a pure substance consisting of two or more elements chemically combined together in definite mass proportions.

In water, for example, the elements hydrogen and oxygen are combined together in the proportion of 1 mass part (e.g. 1 gram) of hydrogen to 8 mass parts (e.g. 8 grams) of oxygen. Similarly, iron ore (iron oxide) contains iron (7 mass parts) and oxygen (3 mass parts). So 10 kilograms of iron ore would yield 7 kilograms of iron.

Wherever this iron ore is found, its elements, iron and oxygen, are always combined in this proportion or ratio. Let us consider one more example. There are several different kinds of sugar compounds, and they all contain the elements carbon (charcoal is impure carbon), hydrogen, and oxygen. One of these compounds, glucose, contains these elements in the mass proportions of 6 (of carbon) to 1 (hydrogen) to 8 (oxygen). Thus, if 15 (6 + 1 + 8) grams of glucose are strongly heated so that all the hydrogen (1 g) and oxygen (8 g) are driven off, a residue of 6 g of black carbon remains. This was the dark substance obtained when you heated sugar in Experiment 4.1.

The atoms in a compound are not all the same kind. They vary in size and mass, depending on which elements they come from. Even to experienced chemists it is something of a wonder that elements can combine together to produce an entirely different substance; that two invisible gases (hydrogen and oxygen), for example, can combine with a black solid (carbon) to form a sweet, white substance (sugar), or that these same two gases can combine to form a liquid (water). The way that elements combine chemically by means of even tinier things than atoms, called electrons, is complicated and something for you to learn later. One can think of everyday examples of things which can combine together rather mysteriously to form different kinds of substances. A loaf of bread is hardly recognizable from its components, flour, water, sugar, and yeast. It is perhaps a little surprising that sugar and butter can be made into toffee and, with milk added, into fudge.

Mixtures

Although most ordinary substances, such as earth, wood, glass, and cloth, are *mixtures* of pure substances (elements and compounds), some are compounds, like water and salt; and a few are elements, such as carbon. Remember that the components of a *mixture* are never in definite mass proportions. For example, powdered sulphur and powdered iron (iron filings) can be mixed together in any

Some pieces of iron ore

Iron ore glittering on the surface of a mine

proportions, such as 4 g iron to 7 g sulphur, or 150 g sulphur to 11 g iron. The elements do not combine together chemically when mixed, and although the mixed powders look greenish in colour you can easily see the yellow sulphur and the silvery iron particles by viewing them through a lens. But if iron and sulphur are heated together in a test tube there is a sudden red glow, showing that they are combining chemically to form a new substance called iron sulphide. It is quite different from either iron or sulphur. And it can be shown that exactly 7 mass parts (e.g. grams) of the iron have combined with exactly 4 mass parts of sulphur.

Experiment 5.1. To show the differences between a compound and a mixture of iron and sulphur

(For older pupils during revision and further studies. DON'T rules (page 147) should be read before doing the experiment.) Make a mixture of approximately 7 g of fine iron filings and 6 g of powdered sulphur in an evaporating basin or beaker. Thoroughly mix the two powders together with a glass rod or spatula. Now divide the mixture into two parts. Put *part A* on to a filter paper so that you can examine the properties of the *mixture* by carrying out the tests described below. Put *part B* into a hard glass test tube so that you can turn this portion of the mixture into a *compound* and carry out the *same* tests on it. It is better to make the compound first, because while it is cooling you can be testing the mixture (part A).

Making the compound of iron and sulphur
Using a test tube holder, gently heat part B in the test tube. You will notice that some of the sulphur vaporizes and con- denses on the cooler part of the test tube. The rest remains as a dark liquid, mixed with the iron. Observe the dark mass all the time until you see the red glow or flash which indicates that the two elements are combining. How much of the sulphur is needed by the iron for the compound to be made? Heat gently for about another minute (about five minutes total time if you do not see the glow) and then place the test tube in a mortar to cool. Now test the mixture (part A).

Testing the mixture
(a) Examine the mixture with a lens. Can you still see the separate little pieces (particles) of iron and sulphur?
(b) Hold a small magnet under the filter paper, and see if you can separate some of the iron particles from the sulphur by moving the magnet about.
(c) Place a little of the mixture in a test tube and add a few cubic centimetres of dilute hydrochloric acid from the bottle on the bench. Which floats, the iron or the sulphur?

A few bubbles of gas will be seen in the liquid if you look carefully. Do they come from the sulphur or the iron? Evidently the acid and one of the elements are causing a chemical reaction which forms a gas. Heat the test tube gently over a Bunsen flame for a few moments to make the reaction go faster and produce more of the gas. *Do not boil the liquid.* As soon as the bubbles of gas are coming steadily and reasonably fast, stop heating and trap the gas in the test tube with your thumb or palm of your hand. After about fifteen seconds, hold the open end of the test tube as close as you can to the Bunsen flame and quickly remove your thumb. What happens? What is the gas? Which two substances reacting together have produced this gas?

(d) If the mixture (iron and sulphur) had consisted of quite different proportions, say 10 g sulphur and 5 g iron, do you think this would have made any difference when you were doing the above tests?

Testing the compound (*iron sulphide*)

First, you will have to get some pieces of the grey compound out of the test tube. Place a bench cloth over the test tube in the mortar, and then carefully break the test tube with the pestle. Remove the cloth, and pick out several pieces of the compound. *Be very careful* not to cut your fingers on thin, sharp pieces of glass. Now carry out the same tests you did with the mixture.

(a) Can you see separate iron and sulphur particles with the lens?

(b) Can you separate the iron from the compound by using a magnet?

(N.B. *Tests (c) and (d) should be done in a fume cupboard.*)

(c) Does one of the elements (iron or sulphur) float and the other sink when you put a piece of the compound in dilute hydrochloric acid? Do bubbles of gas come from the compound more quickly or less quickly than they did from the mixture, or is the rate the same? Do you think it is the same gas? Has it any smell? Does it pop or squeak when the test tube is held to a flame?

(d) Carry out an extra test to see if the gas behaves in any particular way, differently from the other gas. Do test (c) again, making sure that the test tube is very clean, especially at the top. Trap the gas and burn it three or four times. Do you see a flame when it burns? Is anything deposited on the inside of the test tube near the top? If so, what do you think the substance is? Did you get a deposit when the other gas burned? If you are not sure what the substance is, burn the gas several more times until the deposit gets thicker and has a pale colour. Now you will be able to guess what it is. You have probably concluded that the gas is not the same as the first gas, and you will certainly have noticed its foul smell. Its name is hydrogen sulphide. Now write a conclusion to the experiment, pointing out the different properties of (i) a mixture of the elements, iron and sulphur, and (ii) a compound of the same elements, under the following headings: Appearance; Ease of separation of the elements; Reaction with dilute hydrochloric acid. (Note carefully—hydrogen sulphide is a poisonous gas.)

Identification of elements, compounds and mixtures

How can you tell whether a substance is an element, a compound, or a mixture? In Chapter 3 you learned the various methods of separating mixtures into elements and compounds. If you find that you can separate a particular substance into others by any one of these methods it is obviously a mixture.

But to decide whether you have an element or a compound is more difficult. If a substance can be made to decompose, by heating for instance, it is of course a compound. If all methods to decompose it fail, it must be an element, but some of these methods are complicated. At this stage of your chemistry course you are only able to try the heating method, and you have already discovered that some compounds do not decompose when heated. However, by the time you have learned about the appearance and main properties of the common elements you will be able to decide quite confidently that any other pure chemical you see in the laboratory is a compound.

Symbols and formulae

In the list of elements at the end of this chapter you will see that each *element* has a *symbol*, consisting of one or two letters. Note carefully that a single letter, such as 'C' for carbon, is always written as a capital, and that if the symbol consists of two letters, like 'Ca' for calcium, the first letter is a capital. Some of the symbols are based on Latin names. A symbol not only indicates an element, it also means one atom of the element.

The group of letters for a *compound* is called a *formula*, which simply consists of the symbols of the elements in the compound, written close together, to show that the elements are combined chemically. Examples are ZnO for zinc oxide, a compound of zinc and oxygen, and FeS for iron sulphide, a compound of iron and sulphur. These two formulae also show that the compounds contain their elements in the proportion of one atom of one to one atom of the other; for example, in zinc oxide there are equal numbers of zinc and oxygen atoms. But in most compounds the proportions of the elements are different. A well known example is water; its formula, H_2O, means that the elements are combined in the proportion of two atoms of hydrogen to one of oxygen. Here are some formulae of compounds you have already met with.

Table 5.1

Compound	Formula
Copper sulphate	$CuSO_4$
Copper carbonate	$CuCO_3$
Copper oxide	CuO
Carbon dioxide	CO_2
Sodium chloride (salt)	$NaCl$

You will learn more about formulae in Chapter 16.

Questions

1. What is an element? Give two examples.
2. Name 2 metal elements and 2 non-metal elements.
3. What can you say about the tiny particles, or atoms, of which an element is made?
4. 'Elements are the bricks of matter'. What do you think this statement means?
5. Twelve common substances are listed on page 32 which are made of only four different elements. Can you name these elements?
6. Most elements are found combined with other elements in compounds. Can you name 3 elements (a metal, a non-metal solid, and a gas) which are sometimes found in the free state (by themselves)?
7. A compound contains elements. In what way does a compound of elements differ from a mixture of the same elements?
8. Name two common substances which are elements, two which are compounds, and two which are mixtures.
9. List 6 compounds and say which elements each contains.
10. What can you say about the atoms of a compound?
11. How can you tell whether a substance is a mixture or a compound?
12. What is the correct name for the 'abbreviation' of (a) an element, (b) a compound? What are the 'abbreviations' for (c) iron, and (d) copper sulphate?

13. Describe the type and number of atoms in copper carbonate ($CuCO_3$). What else does the symbol of an element mean, in addition to being a 'shorthand' form for the name?
14. Write down the symbols for lead, sodium, zinc, magnesium, and mercury.
15. Write down the names of the elements for which the symbols are K, Al, C, Cl, and N.

Table 5.2 Common elements and their chemical symbols

Name	Chemical symbol
(a) *Metals*	
Aluminium	Al
Calcium	Ca
Copper (Cuprum)	Cu
Iron (Ferrum)	Fe
Lead (Plumbum)	Pb
Magnesium	Mg
Manganese	Mn
Mercury (Hydrargyrum)	Hg
Potassium (Kalium)	K
Silver (Argentum)	Ag
Sodium (Natrium)	Na
Tin (Stannum)	Sn
Zinc	Zn
(b) *Non-metals*	
Bromine (liquid)	Br
Carbon (solid)	C
Chlorine (gas)	Cl
Hydrogen (gas)	H
Iodine (solid)	I
Nitrogen (gas)	N
Oxygen (gas)	O
Phosphorus (solid)	P
Silicon (solid)	Si
Sulphur (solid)	S

Latin names are shown in brackets when symbols have been derived from them.

6 Water as a Hiding Place for Chemicals

Discussion and Investigation

Figure 6.1

It is easier to float in sea water than in fresh water. In the Dead Sea you can float almost on the surface.

Why is sea water more buoyant than fresh water? Do you think river water is pure? What sort of substances might one expect to find in river water? Suppose you wanted to purify some water which you knew to contain chemicals dissolved in it as well as insoluble matter like mud. How would you set about it and what apparatus would you need?

You have probably heard of the word 'solvent', particularly in connection with cleaning. Grease spots on clothing are removed by using a suitable solvent such as white spirit, which is also useful for cleaning your hands when you get oil paint on them. Do you know of any other solvents and what they are used for? Although water does not dissolve grease and paint it is a very good solvent. Write down the names of all the substances you know which dissolve in water (*soluble* substances) and those which are *insoluble*, i.e. do not dissolve.

If you are dissolving something such as salt in water and there is a little left on the bottom which will not dissolve, what can you do to make this last bit dissolve? Is there more than one way of doing this?

Experiment 6.1

Evaporate separately about 10 cm³ of (a) sea water, (b) tap water, (c) distilled water using a watch glass on a steam bath. Compare the amount of residue in each case.

Experiment 6.2. **To compare the solubility of substances at room temperature**

This experiment enables you to find out whether a substance is soluble, moderately soluble, or insoluble. You may also be able to decide whether some soluble substances dissolve more easily than others. Test the following elements and compounds : sodium chloride (salt), sulphur, potassium sulphate, sugar, and copper carbonate. Put 50 cm³ cold water in a beaker and then add 2 g of the substance. Stir the mixture and note roughly how much of the solid dissolves and how quickly.

Experiment 6.3

Repeat Experiment 6.2, but with the following substances : potassium chloride, powdered charcoal (carbon), sodium hydrogen carbonate (sodium bicarbonate), ammonium chloride, and copper oxide.

If you have dissolved as much of a substance as you can by heating the water so that the last little bit has just disappeared, what do you think will happen when the liquid cools?

Experiment 6.4

Heat about a spatula measure of lead chloride and about 20 cm³ of distilled water together in a *large* test tube until the water boils and all the lead chloride dissolves. If it does not quite dissolve add a little more water and boil again. Now cool the test tube under a cold tap, every now and then removing it and looking carefully into the liquid. After a few minutes you will see a rather interesting effect. Can you explain it ? If you miss the effect and find a deposit of lead chloride on the bottom, boil again and repeat the procedure more carefully.

Have you ever travelled over a suspension bridge like the one over the River Severn? You will have been impressed by its length

The Severn bridge

and the clever way in which it has been built. It is suspended or held up by huge steel cables. In chemistry, a *suspension* is a liquid, such as water, in which masses of very small solid particles are suspended by the liquid particles. Muddy water is an example of a suspension. Can you think of other examples? How do you know if a liquid is a solution or a suspension?

Experiment 6.5

Make a suspension by powdering some blue blackboard chalk and shaking it up in water. Filter the blue liquid. Is the filtrate blue? What is the residue on the filter paper?

Explanation and Further Information

Different kinds of water

Although water is a chemical compound (hydrogen oxide) and cannot therefore vary in any way, we talk about sea water, tap water, and distilled water as though there are different kinds of water. The various 'types' of water only differ because of the chemicals contained in them. Sea water contains salt and many other valuable substances such as potassium and magnesium compounds and even gold. River and lake water contains much less of these substances, which are picked up as the water flows along. Certain chemicals are dissolved, and others react chemically with the water and carbon dioxide dissolved in it and so get absorbed by the water. An example of one of these is chalk, which makes water 'hard', i.e. difficult to wash with. Other things, such as fine sand or mud, tiny pieces of plants and leaves, get into the water and form a suspension (see above). So the purification of water involves filtration as well as chemical treatment. Rain water is almost pure water, containing only dissolved gases. The purest of all is freshly distilled water.

Solutions

When a substance is dissolved in water the mixture is called a *solution*. Thus salt dissolved in water is a solution of salt. The substance dissolved is called the *solute* and the liquid in which it is dissolved is the *solvent*, in this case water. Note carefully that a solution is a mixture and not a compound. The particles of solute get in between the solvent particles, but no chemical reaction takes place. Remember then:

$$solution = solvent + solute$$
example: $$salt\ water = water + salt$$

Most solutions that you handle in the laboratory contain solids, like salt. But a solute can be another liquid, such as ethanol, or a gas. River water contains dissolved air, and this is how fish obtain their oxygen.

When a substance dissolves in water its particles become as tiny as the water molecules. This is why a solution is transparent, even when it is coloured, like blue copper sulphate solution. The dissolved particles in a solution go through a filter paper, as you found when you were purifying rock salt.

Water is only one kind of solvent. Some substances, such as sulphur, oil, and iodine, dissolve in it to a very small extent. There are special solvents for them. Sulphur will dissolve in toluene (methyl benzene), for example, and oil or grease in white spirit. Even white spirit will not remove the spray paint used on cars. It needs another kind of solvent. But none of these solvents dissolves such a variety of substances as water, which is sometimes called the 'universal solvent'.

When you have dissolved as much salt as you can in, say, 200 cm³ of water at a particular temperature, you have made a *saturated solution* of salt *at that temperature*. If you heated the water more, the solution would not be saturated because it could then dissolve more salt. In Experiment 6.4 you found that you had to make the water very hot before the lead chloride would dissolve, and when you cooled the solution the lead chloride appeared again as tiny solid particles. When a saturated solution is cooled some of the solute comes out of the solution and the remaining solution is then saturated, with less solute, at a lower temperature.

Solutions are very important in chemistry. In due course you will learn why there are always many solutions of substances in a laboratory. You will discover, for instance, that many solid substances which do not react when mixed together, do so when their solutions are added together, and that chemical reactions between solutions are usually rapid.

Solubility

Elements and compounds differ greatly in their solubility. In Experiments 6.2, 6.3, and 6.4 you found that they can be insoluble, soluble, slightly soluble, etc. You also discovered that most solids dissolve more easily in a hot solvent. Gases are different. They dissolve more easily when the solvent is cold. Ammonia gas, for instance, is so soluble that about 1200 cm³ can dissolve in 1 cm³ of ice-cold water. At room temperature only about 700 cm³ will dissolve in this quantity of water. When you do experiments you should always try to learn whether a substance is soluble or insoluble. *The solubility of a substance is the mass of it which will dissolve in 100 g water at a particular temperature.* For example, the solubility of sodium chloride (salt) is 40 g at 100 °C, which means that 40 g will dissolve in 100 g (also 100 cm³) water at this temperature. What is wrong with the statement that 'the solubility of potassium chlorate is 25 g'?

If a solid is finely divided (powdered) it dissolves more quickly because a larger surface area of the solute is then in contact with the solvent. Stirring increases the rate of dissolution (dissolving) for the same reason since it prevents the solute particles from becoming a mass of sediment and scatters them amongst the solvent particles.

To compare the solubility of different substances it is a good idea to draw a graph (solubility curve) for each substance on the same graph paper. You can then see how solubility varies with the temperature; and you will find that although one substance is more soluble than another at ordinary temperatures, they sometimes switch round at higher temperatures. For example, sodium chloride (salt) is more soluble than potassium nitrate (saltpetre) below about 25 °C but above this temperature potassium nitrate is more soluble than sodium chloride. Figure 6.2 shows some typical solubility curves.

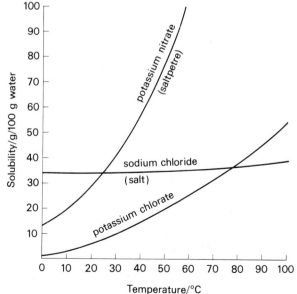

Figure 6.2

Make your own solubility curve for copper sulphate on some graph paper from the following data.

Temperature/°C	0	20	50	75	100
Solubility/g/100 g water	15	22	34	50	74

Suspensions

In Experiments 3.1 (Chapter 3) and 6.5 you separated suspensions easily by filtration. You found that even the smallest of the solid particles would not go through the filter paper. The particles of a suspension, unlike the solute particles in a solution, are large enough to be seen. Since they are heavier than the water particles they eventually settle on the bottom as a *sediment*. This is why it is necessary to stir suspensions before use, examples being whitewash and oil paints.

Emulsions

An emulsion is made when two liquids which are immiscible (do not mix but form separate layers) are shaken together. When cooking oil, for example, is shaken with water a milky-looking liquid is obtained. It consists of minute drops of oil floating in the water. It is like a suspension except that the insoluble particles are liquid instead of solid. Eventually, the dispersed drops of the emulsion come together to form larger and larger drops until the two liquids separate entirely as two layers, as in suspensions. Sometimes certain chemicals, called emulsifying agents, are added to prevent the separation of the two liquids. For instance, soap, which attracts both oil and water, is an emulsifying agent for a cooking oil/water emulsion because it keeps the particles of these two liquids together. Other examples of emulsions are milk (mainly drops of butter fat in water) and mayonnaise (olive oil, vinegar, and egg yolk), both of which contain natural emulsifying agents.

A tug spraying an emulsifier onto an oil slick to break it up

Questions

1. Explain the terms *soluble, solvent, solute.*
2. Give examples of substances that are soluble, moderately soluble, and insoluble in water.
3. What do you think is the main difference between sea water, river water, and rain water, and how could you show this difference by chemical experiment?
4. What is the purest type of water?
5. Why is water sometimes 'hard', i.e. difficult to wash with? How could you make such water 'soft', so that washing with it would be easier?
6. When salt is dissolved in water, is a compound formed? If your answer is yes, name the compound; if no, explain the reasons for your answer.
7. When the blue powder of hydrated copper sulphate is dissolved in water, the powder seems to disappear and the blue solution is transparent. Explain this.
8. Name a solvent other than water, and a substance that will dissolve in it.
9. What is a saturated solution, and what happens when it is cooled? How can you be sure that a solution is saturated?
10. What is the difference between a suspension and a solution? Give examples.
11. What is an emulsion? Give an example.
12. How is the solubility of a substance expressed numerically?
13. Suppose that you are making a solution of potassium chlorate and are using 100 cm³ of water. After 8 g of the chlorate have been added you find that no more will dissolve, the temperature then being 25 °C. You heat the water a little and at 30 °C you find that 10 g will dissolve. If you now allow this solution to cool back to 25 °C, what will happen?
14. If 15 g of a substance dissolve in 300 cm³ of water at 40 °C, what is its solubility?
15. Draw a solubility curve for potassium nitrate from the following data:

Temperature/ °C	10	30	50	70	80
Solubility/g/100 g water	20	45	85	138	170

7 Metal-eaters and Salt-makers

Discussion and Investigation

Have you ever accidentally started to eat your grapefruit without sugar? Some people prefer it this way, but unsweetened grapefruit would probably make you screw up your face. A lemon without sugar is even worse. Suck one in front of a brass band and the music might stop! The sour taste in lemons and grapefruit is due to citric acid, a compound used for making home-made lemonade. Vinegar has a similar sour or sharp taste, for it contains ethanoic (acetic) acid. It hurts quite a lot if you get an acid substance like lemon juice in a cut on your hand. The pain is something like the sting of a bee or wasp. Perhaps bees squirt acid when they sting?

Do you know of any other kinds of acids? You have probably heard of battery acid for cars. Sometimes doctors remove hard skin with an acid. Another kind of acid is used in soldering. You may know the names of these three acids? Unlike lemon juice and vinegar acids, they are very strong and have to be handled carefully. Because they can dissolve most metals it is true to say they are 'metal-eaters'.

Stomach pains and stomach upsets are often caused by indigestion. This is usually due to an excess of acid in the stomach. Perhaps you have experienced this after a birthday party? If so, you were probably given a stomach powder, for example magnesia. Such a substance destroys (neutralizes) acids and is called an *alkali*. Why does toothpaste contain an alkali?

You must now find out for yourself a few things about acids and alkalis. Before you do so, read and heed the following warning:

Be careful with acids and alkalis. When concentrated, they can cause serious injury to the skin, and some give off poisonous fumes when heated. *Never* taste them or any other chemical unless instructed to do so by the teacher, as in Experiment 7.1. (See also the Appendix on page 147 concerning safety precautions.)

Acids

Experiment 7.1

Stir about 5 g citric acid (BP) into about 200 cm³ of water. Wet your finger with the liquid and put a drop on your tongue. Note the sour or sharp taste. Try it again, but this time put a pinch of stomach powder on the acid on your tongue. Does the acid taste disappear?

Experiment 7.2

Put a few cm³ of one of the dilute laboratory acids in a test tube and then add a few drops of blue litmus solution. Note the colour change. This is a safer way than tasting to tell if a liquid is an acid.

Experiment 7.3

Put a few cm³ of one of the dilute laboratory acids in a test tube and add a small piece of magnesium. What happens? Does the acid 'eat' the metal completely?

Experiment 7.4

Repeat Experiment 7.2, and then add to the red liquid a few cm³ of one of the bench alkalis, such as lime water, until the red colour disappears. Has the alkali destroyed the acid? Now you can write down the following information:

Acids, when mixed with blue litmus turn a . . . . . . colour.
Alkalis, ,,   ,,   ,,   red   ,,   ,, ,,. . . . . . colour.

Experiment 7.5

Add about 5 cm³ dilute hydrochloric acid and about 50 cm³ lime water to an evaporating basin and boil to evaporate the liquid. Use a steam bath for the final stages of evaporation to prevent spurting. When a *dry* white powder is obtained, add a little warm water and then taste a drop on your finger. What does the liquid taste like? Has the acid been destroyed?

In Experiment 7.4 you probably used lime water as your alkali. It is a solution of a compound called calcium hydroxide (slaked lime). Why do you think lime is sometimes dug into the soil of gardens and farms? Can you suggest an experiment whereby you could find out whether a sample of soil contains any acid?

Your experiments have shown some of the properties of acids, and you will always remember that acids turn blue litmus red. Litmus can therefore be used to *indicate* whether a liquid is an acid or contains some acid. There are many other *indicators*, substances which change colour when added to acids and alkalis.

Experiment 7.6. To show that acids change the colour of indicators

Add a few drops of each of the following indicators separately to a few cubic centimetres of one of the bench acids. Wash out the test tube each time before an indicator is added to the acid, or use separate test tubes for each of litmus, methyl orange, phenolphthalein, and BDH Universal Indicator.

Make a table in your rough book to show the colours:

	Litmus	*Methyl orange*	*Phenolphthalein*	*BDH Universal Indicator*
Colour with acids				

Acids are corrosive substances and attack skin and clothing. If you get some acid on your hands, wash it off under a tap. Clothing should be treated with a wet sponge or cloth. Concentrated acids are handled only by the teacher. In the following experiment you can find out what happens when dilute acids react with some metals.

Experiment 7.7

Add a small quantity of each of the following metals separately to a few cm³ of dilute sulphuric acid in test tubes, and then repeat with dilute hydrochloric acid: zinc, iron, copper, and magnesium. If the reaction is slow, warm the acid a little in a Bunsen flame, but *do not boil*. You should be able to hold the bottom of the test tube after you have warmed it. If you see bubbles of gas coming from the acid, trap the gas by holding your thumb (or palm of your hand) over the test tube for about fifteen seconds. Now test the gas as follows: hold the mouth of the test tube as near as possible to the flame and remove your thumb. What happens? Which metal gave the most gas?

Which did not give any? The name of the gas is hydrogen, and the test you carried out is always used to identify it. (Note: Use a cork instead of your thumb if the liquid is very corrosive or too hot.)

One of the compounds you heated in Experiment 4.2 was a green substance called copper carbonate. There are many other carbonates, such as magnesium carbonate and lead carbonate. Acids react with carbonates, as you will discover in the next experiment.

Experiment 7.8

Place a spatula measure of any carbonate in a test tube and add a few cm³ of hydrochloric acid. The fizzing (effervescence) shows that a gas is being formed. Test it as you did in Experiment 7.7. Is it the same gas? Fit a cork and delivery tube to the test tube (as in Experiment 4.6), and half fill another test tube with lime water. Now add a little more carbonate and acid to the test tube, replace the cork, and dip the end of the delivery tube into the lime water. What happens when the gas bubbles through the lime water? This effect proves that the gas is carbon dioxide, a compound you will learn about in Chapter 8.

Some salts (page 55) form acids when they are dissolved in water. An example is sodium hydrogen sulphate, which produces sulphuric acid in this way. This salt is therefore one of the chemicals found in most chemistry sets since it is safer to handle than sulphuric acid itself.

Alkalis

Like acids, alkalis are corrosive substances, so if you get some alkali on your hands or clothing take the same action as for acids (page 48). In other ways alkalis are just the opposite to acids, so they neutralize them; this means that the acid and alkali destroy each other in a chemical reaction. Your teacher will probably show you how the strong alkali sodium hydroxide can be neutralized by hydrochloric acid, a reaction which produces a very well known compound.

Solutions of alkalis make indicators change colour, but they are not the same colours as those produced by acids.

Experiment 7.9. To show that alkalis change the colour of indicators

Add a few drops of each of the following indicators separately to a few cm³ of one of the bench alkalis in a test tube. Wash out the test tube each time before an indicator is added to the alkali; or use separate test tubes for each of litmus, methyl orange, phenolphthalein, BDH Universal Indicator.

Combine your results with the acid colours in a table as follows:

	Litmus	Methyl orange	Phenolphthalein	BDH Universal Indicator
Acid colour				
Alkali colour				

Alkalis vary in strength. The strong ones, like sodium hydroxide, are as dangerous as strong acids because they readily attack skin and clothing. There are also certain salts (page 55) which form alkalis when they are dissolved in water. Examples of these are sodium carbonate (washing soda) and sodium hydrogen carbonate (baking soda).

Experiment 7.10. To show that some salts form acids and others form alkalis when dissolved in water

Make solutions of (a) sodium hydrogen sulphate and (b) sodium carbonate (washing soda) by adding a few crystals of each substance separately to about 100 cm³ water in small beakers. Stir both solutions to help the crystals to dissolve. Now add a few drops of Universal Indicator to each solution and note the colour change.

Indicators

Litmus is a purple substance made from certain lichens. Other types of plants can also be used for making indicators. Red cabbage juice and any deep coloured fruit juice are satisfactory. You can try them at home with citric acid and then with a solution of washing soda or baking soda as an alkali. A good indicator is made by extracting the colouring matter from red or pink rose petals.

Demonstration 7.11. To make an indicator from rose petals

Fit a litre flask with a reflux condenser or long glass tube. Place about 200 cm³ industrial methylated spirit (clear) in the flask and fill with red or pink rose petals. Boil the spirit for about twenty minutes until most of the petals have lost their colour. Decant the deep pink solution. Test it with an acid and an alkali to see the colour changes.

A reflux condenser causes the vapour from the liquid to condense and reflux (flow back) into the flask, thus reducing loss of liquid by evaporation and preventing flammable vapour from coming into contact with the Bunsen flame.

Indicators do more than change colour when added to acids and alkalis. They change their colour at a particular strength of acid or alkali. BDH Universal Indicator is a mixture of several indicators, so it changes into different colours depending on the strength of the acid or alkali. This is very useful since it is often necessary to know how strongly acidic or alkaline a substance is. You can watch the colours changing by slowly adding an acid to an alkali mixed with a few drops of BDH Universal Indicator.

Experiment 7.12. To show the colour changes of Universal Indicator

Make a solution of citric acid by stirring about 2 g of the acid into about 300 cm³ water. Place about 25 cm³ lime water in a small beaker and add a few drops of the indicator. Now add the acid, very slowly, using a teat pipette, and stirring the liquid with a glass rod. Although quite a lot of acid has to be added to produce the colours, remember that it is the *last few drops* that cause one colour to change to another. If you miss a colour by adding the acid too quickly, pour a little more lime water into the beaker and start again. Make a list of the colours you obtain, then see if you can answer these questions.
(a) What is the colour of the indicator when mixed with the pure alkali?
(b) What was the next colour you got when the alkali had been weakened (some of it had been neutralized) by adding a little acid?
(c) What was the final colour obtained when a lot of acid had been added and there was no further change?
(d) Between the stages in (b) and (c) there must have been a point when just sufficient acid had been added for exact neutralization when there was no acid or alkali in the

beaker. What do you think is the colour representing this stage ?
(e) Is there another colour which comes when there is just a little acid left over after neutralization, i.e. when the acid is weak ?

Salts

In Experiment 7.5 you obtained a white powder when you neutralized hydrochloric acid with lime water (calcium hydroxide solution). You may have seen a demonstration to show the neutralization of this acid by sodium hydroxide solution. If this is done in special apparatus accurately, so that there is no acid or alkali left over, salt water is produced and from it pure salt (sodium chloride) can be obtained by evaporation. The white powder you made is another kind of salt (not edible) called calcium chloride. Whenever an acid and alkali neutralize each other a salt and water are formed. This word equation will help you to remember:

$$ACID \quad + ALKALI \rightarrow SALT \quad + WATER$$

example: hydrochloric + sodium   → sodium + water
                      acid   hydroxide   chloride

There are a great many different kinds of salts, depending on the acid and alkali from which they are derived. Sodium sulphate, for example, is the salt produced (with water) when sulphuric acid is completely neutralized by sodium hydroxide.

Here are some word equations showing how some salts can be made. Some of the equations have been only half written so that you can complete them in your note book.

Table 7.1

Acid	Alkali	Salt	Water
Hydrochloric acid (Hydrogen chloride solution)	+ Sodium hydroxide	→ Sodium chloride	+ Water
Hydrochloric acid	+ Calcium hydroxide	→	+ Water
Hydrochloric acid	+ Magnesium hydroxide →		+ Water
Sulphuric acid (Hydrogen sulphate solution)	+ Sodium hydroxide	→ Sodium sulphate	+ Water
Sulphuric acid	+ Potassium hydroxide →		+ Water
Nitric acid (Hydrogen nitrate solution)	+ Sodium hydroxide	→ Sodium nitrate	+ Water
Nitric acid	+ Calcium hydroxide	→	+ Water

You will have noticed that each salt contains a metal and part of an acid. Study the word equations carefully and then try to write down a good definition of a salt.

Explanation and Further Information

Acids

An acid is a compound which, when dissolved in water, has a sour or sharp taste and changes the colour of indicators, e.g. blue litmus to red litmus. Acids are corrosive substances and attack most metals, the products depending on the acid used and on the particular metal. When acids are dilute their reaction with metals usually produces a salt and hydrogen. Here is an example:

Hydrochloric acid + zinc → zinc chloride + hydrogen

When an acid neutralizes an alkali, a salt and water are always formed:

Sulphuric acid + sodium hydroxide → sodium sulphate + water

With carbonates, acids produce the gas carbon dioxide as well as a salt and water:

Nitric acid + copper carbonate → copper nitrate + water
+ carbon dioxide

All acids contain hydrogen.

Strong and weak acids

Acids which are found in plants and animals are fairly weak. They are called *organic acids* because they have come from living organisms. Examples are citric acid (found in citrus fruits), ethanoic (acetic) acid (in sour wine or vinegar), lactic acid (sour milk), methanoic (formic) acid (stings of ants and bees).

The strong acids are classified as *mineral acids* because they are produced from mineral substances such as sulphur and salt. These are the dangerous acids, especially when they are concentrated, i.e. contain little water. For most purposes in the laboratory they are mixed with water and are then called *dilute acids*. These solutions of acids are much safer to use, but can still damage skin and clothing.

Uses of acids

Here is a list of some of the better known acids and their main uses.

Table 7.2

Acid	Some uses
Mineral Acids	
Sulphuric acid	Car batteries, fire extinguishers, oil refining, making chemicals.
Hydrochloric acid	Rust removing, soldering, making chemicals.
Nitric acid	Making explosives, spray paints, fertilizers, and various chemicals.
(Note These are the laboratory bench acids.)	
Organic Acids	
Citric acid	The acid in citrus fruit used in wine-making, and home-made lemonade.
Ethanoic (acetic) acid	The acid in vinegar. Used in making paints and textiles.
Oxalic acid	A poisonous acid found in certain plants. Used for removing rust and ink stains.

For the colour of common indicators with acids see p. 56.

Alkalis

An alkali is a compound which neutralizes an acid and changes the colour of indicators, e.g. red litmus to blue litmus. You have already learned about neutralization of acids by alkalis. Alkalis are

Table 7.3

Name	Some uses
Potassium hydroxide (strong)	Making soft soap
Sodium hydroxide (strong)	Soap-making, petroleum refining, paint and grease removing
Calcium hydroxide (moderate)	As 'lime' for neutralizing soil acids
Ammonia solution (moderate)	Household cleaner
Magnesium hydroxide (weak)	Medicinal alkali, toothpaste

For the colours of common indicators with alkalis see p. 56.

corrosive like acids, and the stronger ones attack paint, grease, and some metals. They are therefore useful as paint removers and for cleaning greasy surfaces (e.g. ovens). The common alkalis are given in Table 7.3.

Salts

A salt is the compound formed when the hydrogen of an acid is replaced by a metal. When magnesium, for example, reacts with sulphuric acid, it replaces the hydrogen in the acid, forming the salt magnesium sulphate, and the hydrogen is set free:

magnesium + sulphuric acid → magnesium sulphate + hydrogen

Some acids can form more than one salt with a metal. This happens when only part of the hydrogen is replaced. A hydrogen salt is then formed. For example, sulphuric acid can form normal sodium sulphate (when all the hydrogen is replaced) and sodium hydrogen sulphate (page 50) when only part of the hydrogen is replaced by the sodium.

Salts are evidently named from the acids which produce them. Thus, sodium sulphate comes from sulphuric acid (hydrogen sulphate solution), copper chloride from hydrochloric acid (hydrogen chloride solution), calcium nitrate from nitric acid (hydrogen nitrate solution), and lead carbonate from carbonic acid (hydrogen carbonate solution).

Salts vary considerably in solubility. When they are soluble their solutions may be neutral (sodium chloride, for example), acidic as in the case of sodium hydrogen sulphate, or alkaline (e.g. sodium carbonate).

Table 7.4

Salt	Some uses
Copper sulphate	Garden insecticide and fungicide, stopping footrot in sheep, copper plating
Potassium nitrate	Making gun powder and fuses, soil fertilizer
Calcium chloride	To melt ice on roads, refrigeration plants
Zinc chloride	As a flux in soldering
Sodium chloride	Making chlorine gas, sodium hydroxide, and many other chemicals

Some salts have important uses. Table 7.4 gives a list of some of the better known ones.

Indicators

The colours of the common indicators in acid and alkali solutions are as follows:

	Litmus	Methyl orange	Phenolphthalein	BDH Universal Indicator
Acid	red	red	colourless	red (but see below)
Alkali	blue	orange	pink	violet (but see below)

Note carefully that indicators change colour only when in contact with solutions of substances in water.

The colours of BDH Universal Indicator— pH numbers

In Experiment 7.12 the starting colour, when the indicator was added to the pure alkali, was violet. As the acid is slowly added it gradually neutralizes and therefore weakens the alkali. So the colour changes to blue and then to greeny blue. When the neutralization is complete the colour becomes green (neutral); there is now no acid or alkali, just a salt and water. This is also the colour when the indicator is added to pure water. As more acid is added the solution becomes first a weak acid (yellow) and then gets stronger until the colour red is reached. The different colours of Universal Indicator correspond roughly to what are called the *pH numbers* of acids and alkalis. These numbers tell the strength or degree of acidity and alkalinity more accurately.

Low numbers indicate strong acids and high numbers mean strong alkalis. As the acid numbers get larger the acid gets weaker. pH 7 is a neutral solution (e.g. when an acid is neutralized by an alkali). As the alkali numbers (pH 8 onward) become larger the alkali gets stronger. Figure 7.1 makes all this clear. Note carefully that the concentration of an acid or alkali is not necessarily an indication of its strength (acidity and alkalinity). A concentrated solution of citric acid may be weaker than a dilute solution of a mineral acid. Calcium hydroxide solution of maximum concentration is less alkaline than most dilute solutions of sodium hydroxide.

Colour	red	orange	yellow	green	blue	indigo	violet
pH number	4	5	6	7	8	9	10

ACIDS   N   ALKALIS
E
U
◄——strong   T   ◄——weak
R
weak——► A   strong——►
L

Figure 7.1

You can find the approximate pH number of common substances at home, such as white vinegar, lemon juice, solutions of washing soda, soap, baking soda, etc. by adding Universal Indicator to them and noting the colour.

Questions

1. What is an acid? Name two examples.
2. How do organic acids and mineral acids differ? Give an example of each.
3. Describe, with examples, three different properties of acids.
4. Name two mineral and two organic acids and mention their uses.
5. What are the products when (a) an acid reacts with a carbonate? (b) hydrochloric acid reacts with zinc?
6. Describe two properties of an alkali and name two of these compounds.
7. What are the products when an alkali neutralizes an acid?
8. What is the importance of alkalis in connection with health and with gardening?
9. What is an indicator? Give two examples.
10. Define a salt and give two examples of salts.
11. Write a word equation to show how sulphuric acid reacts with sodium hydroxide.
12. Write a word equation to show how the salt calcium chloride can be made from the neutralization of hydrochloric acid.
13. What are the colours of litmus, methyl orange, and phenolphthalein when added to (a) an acid and (b) an alkali?
14. What are the colours of Universal Indicator when added to (a) a weak acid, (b) a strong alkali, and (c) pure water?
15. A solution, X, had a pH number of 4. A pupil added another solution, Y, to it until the mixture of X and Y had a pH number of 7. Describe solution X, solution Y, and the mixture of X and Y.

8 Gases Galore

Discussion and Investigation

In Chapter 4 you found that most metals, when heated in air, gain in mass by absorbing air or something in the air, and form powders of various colours; you learned about the white, black, yellow, and red powders which are formed by magnesium, copper, sodium, and mercury respectively. Your discovery was a very important one which eluded the early chemists for a very long time. They thought that metals *lost* something when they were heated and that the powders were a sort of ash, like wood ash.

These ideas were eventually proved wrong by two very famous chemists, Joseph Priestley, an Englishman, and the French scientist, Antoine Lavoisier. Both carried out experiments with metals and air in addition to a great deal of other scientific investigation. Priestley was particularly interested in gases; and used a large 'burning glass' (lens) to heat various chemicals. One day he heated the red powder obtained by heating mercury—the liquid metal used

Joseph Priestley 1733–1804

Antoine Lavoisier (1743–1794) and his wife

in thermometers. Collecting the gas which was given off, he experimented with it and found that things burned very brightly in it. He put a mouse in a jar of the gas and it became very lively. He therefore called the gas 'active air'. If you are not old enough to do the following experiment your teacher will probably demonstrate it to you. It must be done in an efficient fume cupboard because *mercury is a very poisonous substance*. For this reason only a small amount of the red powder is heated and will not produce enough oxygen for the mouse!

Experiment 8.1

Place one spatula measure of the powder (now called mercury oxide) in a test tube fitted with a cork and delivery tube. Fill a large beaker or bowl with water, and place in it an inverted test tube of water. The end of the delivery tube will enter the test tube at a later stage (Figure 8.1). Clamp the first test tube in a retort stand, and heat it gently at first, then more strongly, keeping the flame at the bottom of the test tube. When bubbles of air (from the test tube) are seen in the water, wait for about ten seconds and then place the test tube of water over the end of the delivery tube. The air in the first test tube has now been driven out, and bubbles of a gas will be seen rising in the second test tube and displacing the water. When the test tube is full of gas, remove it from the delivery tube, and lean it against the side of the beaker. Take the delivery tube out of the water and stop heating. If you do not do this atmospheric pressure may push the water back into the hot test tube and cause it to crack. Now take a splint of wood and

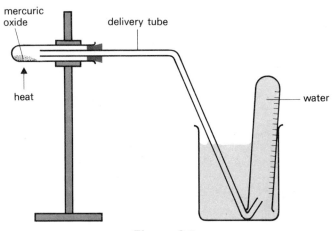

Figure 8.1

light it. Blow it out so that it just glows red. When you have practised this, carefully remove the test tube from the beaker, holding your thumb (or a cork) over the end. Light the splint again, blow it out, and quickly insert the glowing end into the test tube of gas. What happens? Did you find that Priestley was right? You will notice a grey 'collar' of mercury about half way up the first test tube. When this test tube is cool, scrape out the mercury with a spatula into an evaporating basin, being careful to hold the test tube horizontally to prevent the powder coming out. You will probably obtain enough mercury to form a little silver ball which runs about in the basin, hence the old name of 'quicksilver' for mercury.

After Priestley and Lavoisier had met in France to discuss their work, Lavoisier, who already knew that metals *gain* in mass when heated, decided that the new gas Priestley had made was the gas in air which combines with metals when they are heated. So the air must contain more than one gas, he concluded. To prove his theory he carried out the following experiment with the utmost care.

Lavoisier's Experiment

He heated mercury in a retort over a charcoal furnace for several days. The end of the retort entered a measuring cylinder supported in a trough of mercury and containing a known volume of air. He found that the mercury rose up into the cylinder a little way, indicating that the volume of air had

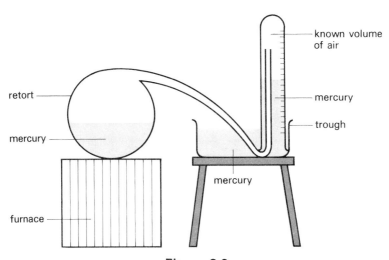

Figure 8.2

decreased. When there was no more change in this volume he allowed the apparatus to cool and then carefully removed all the red powder which had formed on the mercury surface in the retort. He then did your experiment (Experiment 8.1) and found, as he had expected, that the volume of gas collected was exactly the same as the decrease in volume of the air in his first experiment.

Lavoisier tested the gas and confirmed all that Priestley had reported about it. Clearly the part of the air which had originally combined with the mercury to form the red powder was the same gas liberated by the powder when heated.

In various tests with the gas Lavoisier found that non-metal elements, such as sulphur, burned in it to form gases which dissolved in water and formed acids. So he called the gas *oxygen*, which is derived from two Greek words meaning acid producer. He did not then know that some acids do not contain oxygen.

Lavoisier also tested the 'inactive' gas remaining in the measuring cylinder after the first experiment. Nothing would burn in it, i.e. it would *not support combustion*, and a mouse died when put in it. Do you know the name of this inactive gas?

Oxygen is not the only gas which will relight a glowing splint of wood, and in order to be sure that it is indeed the same gas which combines with various metals when they are heated, additional tests should be done. One way of doing this is to heat two or more powders such as mercury oxide and lead oxide, and test the gases collected, but in another way. Gases have different *densities*. The density of a gas is the mass in grams of one litre. However, it is not necessary to weigh one litre of the gas, and this is more difficult than you might think. Can you suggest another way before you read on? All that has to be done is to make the gas (oxygen) by heating the two powders, note the loss in mass of each powder (i.e. the mass of oxygen evolved) and measure the volume of oxygen collected in each case. Thus, if it was found that the oxygen weighed 0.13 g and had a volume of 100 cm³, then its density would be 1.3 g per litre. The oxygen from both powders should have the same density, although experimental results may well differ slightly. You can now try this experiment for yourself.

Experiment 8.2. **To find the density of oxygen at room temperature and pressure**

Place about 8 g (measured roughly on a lever arm balance) well-dried red lead oxide in a dry test tube and weigh on a chemical balance to the nearest milligram. Then fit a cork and delivery tube to the test tube, so that the oxygen can be collected *over water* as in Experiment 8.1, except that a *gas jar* should be used in place of a test tube. The gas jar of water can be supported on a beehive shelf or clamped in a retort stand. Heat the lead oxide until no more oxygen is evolved and disconnect the delivery tube as in Experiment 8.1 as soon as heating is stopped. Now close the open end of the gas jar with a greased glass cover, and place it on the bench. The gas jar will contain some water, and the space above will be oxygen. To find the volume of this oxygen, fill a measuring cylinder with water exactly to its zero mark. Now pour the water from it into the gas jar until the jar is exactly full. The volume of water poured from the cylinder is the volume of oxygen evolved from the red lead oxide. By now the test tube containing the oxide should be cool. Weigh it carefully as before. The difference in the two masses is the mass of oxygen given off. Now you can calculate the density of oxygen as already described. It will be between 1.27 g and 1.42 g per litre, depending on the temperature and pressure of the air.

You have made oxygen and tested it in two ways. What properties of the gas have been shown during your experiments? Is it soluble or insoluble in water? Has it a colour? List all the properties you can think of. You may have learned about oxygen in biology. How do plants and animals use it? Are there other uses of oxygen? How is it made on a large scale? If you have forgotten turn back to Chapter 3.

Composition of the air

What other gases are there in the air? Make a list, and state how you think some of them get into the air. How do you think town air

The Home Office in 1964

The Home Office in 1972 after cleaning

might differ from country air? We hear and read a lot about pollution. What are the main things that pollute the air? What was the method, described in Chapter 3, for removing carbon dioxide from the air?

Can you suggest an experiment to find out the proportion of oxygen in the air? Discuss this together before you read on. Do not use Lavoisier's method because mercury is expensive and, as you know, poisonous.

Now try the following experiment.

Experiment 8.3. To find the approximate proportion of oxygen in the air—iron filings method

Add a few drops of water to the bottom of a measuring cylinder and shake so that the glass is wet. Now add about two spatula measures of iron filings and shake the cylinder a little to make the iron filings stick to the bottom. Invert the cylinder in a large beaker or deep bowl half full of water, having inserted a length of flexible tubing into it to allow some air to escape. This makes the water levels inside and outside the cylinder equal, so the air inside the cylinder is at atmospheric pressure. Note the exact level of water in the cylinder and leave for several days. The rusting of the iron will slowly remove the oxygen, its place being taken by water forced in by atmospheric pressure. When the water level inside the jar does not climb any higher, pour some water into the beaker to equalize the levels, and therefore the pressures, inside and outside the cylinder. Now read the new level of water in the cylinder. The decrease in the original volume of air in the cylinder is the amount of oxygen it contained. From this you can work out the proportion of oxygen in air as about one fifth. Place a cork in the cylinder while it is still in the water and then put the cylinder on the bench. It contains what Priestley called 'inactive air' because he found that the gas would not support combustion. Try this for yourself by inverting the cylinder, removing the cork, and thrusting a lighted taper or splint into it.

Experiment 8.4. To find the proportion of oxygen in the air

Set up the apparatus as in Figure 8.3.

The syringes are connected by about 20 cm of hard glass tubing which contains copper powder. Place a glass wool plug on each side of the copper so that powder is not blown into the syringes. The connections between tube and syringes should be short lengths of pressure tubing. One syringe should contain no air and the other should contain a known volume, say 50 cm^3. Heat the copper powder, gently at first, then

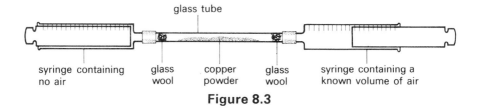

Figure 8.3

strongly, with the Bunsen burner stationary under the glass tube. Now move the syringe plungers backwards and forwards for two or three minutes so that the heated copper combines with the oxygen. Allow the apparatus to cool, and note the new volume of air in the syringe. Heat again and cool as before, until the volume of air is constant. Now calculate the percentage of oxygen in air.

That air contains carbon dioxide can be shown by bubbling air through lime water in the apparatus shown in Figure 8.4. The lime water turns milky, and this is the identification test for carbon dioxide. In clean air this may take some time, and you have to watch carefully because the lime water goes clear again quite quickly.

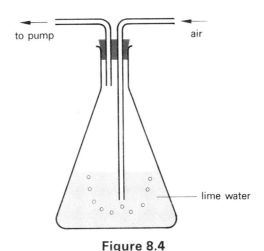

Figure 8.4

Here is an experiment to enable you to prepare carbon dioxide and examine some of its properties.

Experiment 8.5

Assemble the apparatus as in Figure 8.5.

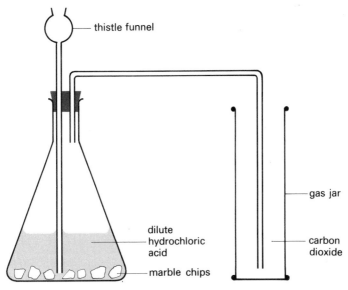

Figure 8.5

Cover the bottom of the conical flask with marble chips. Arrange for the delivery tube to reach nearly to the bottom of the gas jar. Pour dilute hydrochloric acid down the thistle funnel until the level of the acid is above the end of the funnel. (Why is this necessary?) The gas fills the gas jar from the bottom, forcing the air upwards and out of the jar. This method is called collecting a gas by the *upward displacement* of air. You can tell when the jar is full by holding a lighted splint at the mouth. It will quickly go out if the jar is full of gas.

Collect three jars of the gas, placing glass covers on them as soon as they are full. If the evolution of gas slows down, add more acid, pouring out some liquid from the flask to make room if necessary.

(a) *To find out whether the gas is denser than air*
See if you can 'pour' the gas from one of the full jars into a second jar containing a little lime water, then cover the second jar and shake. How will you know if you have succeeded?

(b) *To see if carbon dioxide puts out fires*
Light a candle fixed to a gas jar cover and then pour the gas on to the fire from directly overhead, holding the gas jar close.

(c) *To investigate the nature of the compound formed when carbon dioxide reacts with water*
Pour a little water into one of the jars of gas, replace the cover,

and shake vigorously to make the gas dissolve. Now add a few drops of BDH Universal Indicator. What type of compound was formed by the gas and water?

Can you think of a way to show that air contains water vapour? Water vapour condenses readily on a cold surface. You may have noticed this when you have a glass of cold water or lemonade in a warm room. If a glass is filled with pieces of ice and placed on a saucer in a warm room, the water vapour in the air condenses on the outside of the glass and runs down into the saucer. You can note the effect of adding anhydrous copper sulphate to the liquid, as in previous experiments.

Condensation on a window

Further investigation into the properties of oxygen

You saw some oxygen being made by heating mercury oxide. This experiment demonstrated the following properties of oxygen.

(a) It is a colourless, odourless gas.
(b) It is only slightly soluble in water; in fact it appeared to be insoluble.
(c) It allows things to burn in it, i.e. it *supports combustion*.
(d) Substances burn better in it than in air. (Why?)

In the next experiments you are going to see what happens when metals and non-metals burn in oxygen.

Experiment 8.6. To prepare oxygen and study some of its properties

The apparatus is shown in Figure 8.6. Large test tubes, requiring small combustion spoons, may be substituted for gas jars if preferred. Cover the bottom of a conical flask with granules of manganese dioxide and fit a two-hole bung with a thistle or tap funnel and delivery tube. The end of the delivery tube enters the beehive shelf on the bottom of the trough or bowl of water. Fill four gas jars with water and invert them in the trough. Now add enough hydrogen peroxide (twenty volume strength) to the flask to produce a steady stream of oxygen. If a thistle funnel is used the level of liquid must be above the end of the funnel. (Why?) When the air has been displaced from the flask, put the first gas jar on the beehive shelf and start collecting the oxygen. When all four gas jars are full of oxygen, each is removed in turn for the experiments below. This is done by sliding a glass cover under the open end of the jar, withdrawing the jar from the water, and placing it on the bench with the cover firmly in position so that no gas can escape. Adjust a combustion spoon so that the metal lid will cover the gas jar when the spoon itself is a little more than half way down the jar. Now you are ready to burn the first element.

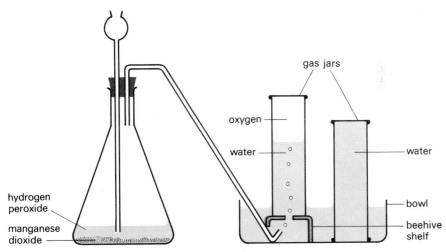

Figure 8.6

1. *Sulphur*
If the laboratory is not well ventilated it is preferable to burn this element in a fume cupboard. Half fill the combustion spoon with powdered sulphur, hold it in the Bunsen flame until it ignites, and then plunge it into the gas jar of oxygen. Note the difference between the way the sulphur burns in air and in oxygen. Has a gas been formed by the burning? When the flame has gone out, remove the combustion spoon, quickly add a little water to the jar, and place a glass cover over it. Now shake the jar to make the gas dissolve in the water. Can you explain how the gas was formed when the sulphur burned?

When the gas dissolves in water, some of it reacts chemically with the water, forming another compound. Add some blue litmus solution to the water. What type of compound has been formed?

2. *Carbon*

Use the second gas jar of oxygen to burn small pieces of charcoal, following the same procedure as for the sulphur. Heat the charcoal until it is red hot, then plunge it quickly into the oxygen. The carbon burns brightly, forming an invisible gas. Shake the gas with some water for about a minute. Pour some of the water into a little lime water in a test tube. What is the gas? Add some blue litmus solution to the rest of the water. What type of compound has been formed by the water reacting with some of the gas?

3. *Iron*

Add half a test tube of water to the gas jar and quickly replace the cover. This is to catch hot sparks of the burning iron and prevent cracking of the gas jar. Twist a long and slender piece of steel wool round the combustion spoon, hold it in the flame, and as soon as sparks are coming well, put the spoon quickly in the oxygen. When the burning is over, examine the hard piece of compound formed. Try to dissolve it in water. What do you think the compound is?

4. *Magnesium*

Twist a few centimetres of magnesium ribbon round the combustion spoon, ignite it and quickly place it in the jar of oxygen. **Do not look at the dazzling flame.** It can damage the eyes. When the magnesium has burned, shake off any loose white powder into the gas jar. What is the white powder? Add a little warm water, and shake vigorously. A little of the powder will dissolve. Now add a few drops of phenolphthalein. A pale pink colour indicates the presence of a weak alkali. What is the alkali? How has it been made? The liquid may have to stand for a few minutes before the colour shows.

Write a word equation for the burning of the magnesium.

Write a conclusion to your experiment concerning the effects of burning

        (a) a non-metal element;
        (b) a light, active metal (magnesium);
        (c) a heavy metal (iron).

Check your findings later by reading page 77.

Experiment 8.7. **To burn sodium, calcium, and zinc in oxygen and test the products**

(For more experienced pupils only.)

Be very careful when you handle sodium. Do not touch it with your hands, and keep it well away from water, with which it reacts vigorously and sometimes explosively. The sodium should be wiped with filter paper to absorb the oil, and the calcium should be rubbed with emery paper to remove any oxide layer.

Put each metal on a clean brick or fireclay crucible lid and play the Bunsen flame and a jet of oxygen (from a cylinder or aspirator) on to it. A piece of sodium the size of half a small pea is satisfactory. A few turnings of calcium are needed, and the zinc should be in the form of a powder. The typical flame of sodium burning soon appears. What is its colour? The calcium takes much longer to burn and strong heating is necessary. You will have to watch carefully to see its colour. The zinc also burns with a coloured flame, and needs quite a lot of heating. When the burning is complete, wash the powders into beakers containing a little water and test each solution with BDH Universal Indicator. Describe the reactions which took place and the compounds formed. Write word equations of these reactions.

Hydrogen

Is hydrogen one of the gases in the air? What compounds contain hydrogen? Can you remember how hydrogen was produced and tested in one of your experiments?

It is the least dense gas and was once used for filling airships. Why was this found to be too dangerous? Do you know the name of the gas now used?

Sometimes, at fairs, you can buy balloons filled with hydrogen. When released they go up into the air a long way. Why do they not continue to ascend?

Do you know of any other important uses of hydrogen?

Can you suggest how it is made on a large scale? In the laboratory it can be made in several ways, but the usual method is the reaction of dilute hydrochloric acid with the metal, zinc. The zinc displaces the hydrogen from the acid. In the following experiment you will find out some of the properties of hydrogen.

Experiment 8.8

Cover the bottom of a conical flask with granulated zinc and fit a two-hole bung with a thistle or tap funnel and delivery tube. Complete the apparatus as in Experiment 8.6 for collecting the hydrogen over water. Now pour the dilute hydrochloric acid down the funnel until the acid level is above the end of the funnel. When the air has been displaced from the flask, place the first gas jar on the beehive shelf. Collect two gas jars and one large test tube of hydrogen, being careful that they do not fall over in the trough of water. Remove the conical flask to a safe distance and light the Bunsen burner. Slide a greased cover over one of the jars, take it out of the water, and place it on the bench.

1. *To find out what happens when hydrogen, mixed with a little air, is ignited*
Light a splint of wood, remove the cover from the jar of hydrogen, and hold the flame to the top of the jar. The effect obtained is used as an identification test for hydrogen.

2. *To see whether hydrogen is less dense than air*
Remove the second jar of hydrogen from the bowl of water, hold an empty gas jar about 3 cm above it, and take the cover off the full jar. Does the gas rise into the top jar? How can you tell?

A hydrogen balloon setting off to cross the English Channel

3. *To discover whether hydrogen* (a) *burns, and* (b) *supports combustion*
For this experiment you use the test tube of hydrogen and need a lighted wax taper or candle. Place a cork loosely in the test tube and remove it from the bowl. Hold the tube inverted, remove the cork, and quickly push the taper well up into the tube. What happens? Does the taper continue to burn? Withdraw the taper as soon as you have pushed it up. What happens now? Put the taper back into the test tube and take it out again. Describe all that you observe and explain it.

In Experiments 8.6 and 8.7 you found that when elements burn they combine with oxygen to form compounds such as carbon dioxide and magnesium oxide. These compounds are called *oxides*. What happens when hydrogen burns? Does it too form an oxide?

Demonstration 8.9. To burn hydrogen and examine the product formed
Safety precautions
A safety screen should be used. In order to prevent an explosion it is essential to ensure that the apparatus contains no air before the jet of hydrogen is lighted. This means that time must be allowed for the hydrogen (preferably from a cylinder) to replace completely the air in the apparatus. Also, a steady

stream of the gas should be maintained throughout the experiment; for this reason it is advisable to add a little concentrated acid to the dilute acid in the tap funnel, assuming the hydrogen is being obtained as in Figure 8.7, and it is necessary that some acid remains in the tap funnel during the experiment in order to ensure that air does not enter the apparatus. All corks should fit well, and a check should be made that gas can easily pass through the anhydrous calcium chloride in the U tube. *From whatever source, the hydrogen must be tested before being ignited*, to ensure that it has not been mixed with air. This can easily be done by collecting samples of the gas over water in large test tubes and igniting them in the inverted test tubes. The gas should burn quietly after the initial 'pop'.

Method
Burn the gas until one or two drops of a colourless liquid drop into the watch glass. Then blow out the flame and disconnect the reaction flask. Add a spatula measure of anhydrous copper sulphate to the liquid. What do you think the liquid is ? How was it formed ? Should other tests be done to identify it ? Why was a cold surface held against the hydrogen flame ? Why was the hydrogen passed through the drying agent, anhydrous calcium chloride ?

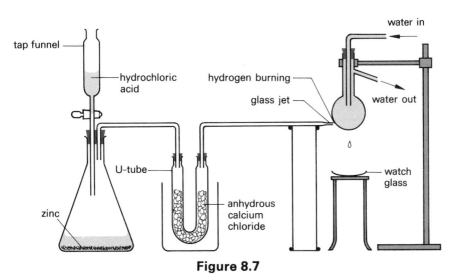

Figure 8.7

Evidently hydrogen combines easily with the oxygen in the air. Is it chemically strong enough to take oxygen out of *compounds*, such as oxides?

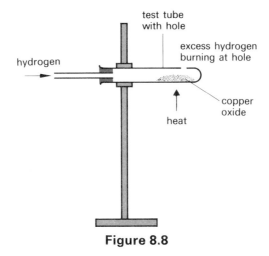

Figure 8.8

Demonstration 8.10. **To find out whether hydrogen can take the oxygen out of copper oxide**

The apparatus used is shown in Figure 8.8. A test tube with a hole in the bottom is used to hold the copper oxide. *The same precautions* must be observed as in Demonstration 8.9, to ensure that there is no air in the apparatus before the oxide is heated. Pass the hydrogen over the heated oxide for about fifteen minutes, allowing the excess to burn at the hole in the test tube. Stop heating, but let the hydrogen continue to flow for a few minutes while the test tube is cooling. Examine the test tube. What is the pinkish substance sticking to the glass? How has it been produced? Has any other product been formed? Write a word equation for the reaction.

Explanation and Further Information

By investigating for yourself, as the early chemists did, you have managed to find out a great deal about the gases of the air and hydrogen.

Air

Air is a mixture of nitrogen (the main, inactive part), oxygen, carbon dioxide, water vapour, and a very small proportion of what are called the noble gases (see page 73).

The water vapour varies a great deal. On a hot, steamy day there is obviously quite a lot of it. When it is very cold and water freezes into snow and ice, there is less. Carbon dioxide is only 0.03% by volume of the air, but it is vitally important. Why?

Although the composition of air can vary very slightly in different places in the world, dry air nearly always contains the following proportions of gases by volume: nitrogen 78%, oxygen 21%, carbon dioxide and noble gases 1%. This is clean air, as found in country districts. Town air is polluted with other substances such as smoke particles, extra carbon dioxide (from fires, factory chimneys, vehicle exhaust gases, etc.), and other gases such as sulphur dioxide (see Experiment 8.6) and the poisonous carbon monoxide, which is emitted in small quantities from vehicle engines.

The gases of the air can be separated by the fractional distillation of liquid air. The air is highly compressed and made extremely cold. so that it turns into a liquid at the very low temperature of $-200\,°C$, When this liquid is distilled, nitrogen boils off first, followed by oxygen. The nitrogen is mixed with the noble gases (see below) and this mixture has to be liquefied and fractionally distilled again to separate the gases.

Nitrogen

Nitrogen, Priestley's 'inactive air', does not support combustion or combine easily with other elements. Yet it is an element in many important and valuable compounds. Plants need nitrogen and obtain it mainly by absorbing nitrogen compounds (e.g. nitrates) from the soil. Nitrogen is used in large quantities for making the gas ammonia, a compound of nitrogen and hydrogen. The ammonia is used for making nitric acid and other chemicals, especially soil fertilizers, and in the production of household cleaning liquids. Proteins, an essential item in our diet, contain nitrogen. We obtain them by eating plant and animal proteins, but our bodies cannot absorb nitrogen gas and turn it into proteins.

Noble gases

The noble gases in the air are helium, argon, neon, krypton, and xenon. They were given their general name for the same reason

Neon lights at Piccadilly Circus

A US Navy helium-filled airship

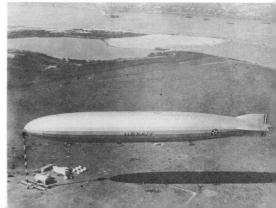

that gold and platinum are called the noble metals, because these elements are reluctant to combine with others, remaining superiorly alone. Gold and platinum do not tarnish as other metals do, i.e. they do not form coatings of oxides or other compounds. Helium is used in airships and observation balloons, and in deep-sea diving equipment. Argon is the gas which is used to fill ordinary electric light bulbs. It does not react with the metal filament in the bulb. Coloured advertising signs contain neon. Krypton and xenon are used for making special flash tubes for taking high speed photographs. The flash lasts only a fifty thousandth of a second.

Carbon dioxide

This gas comes into daily life a great deal. It is the gas in mineral waters ('fizzy drinks') and is used in fire extinguishers. Washing soda and white lead for paints are also made from it. Carbon dioxide plays a vital part in plant and animal life. Plants take in

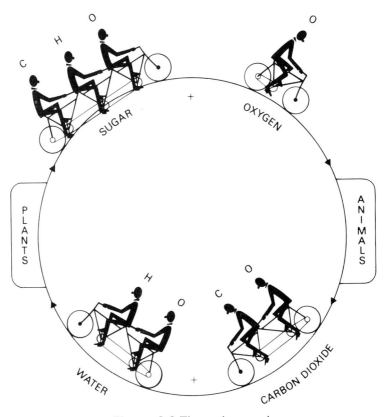

Figure 8.9 The carbon cycle

carbon dioxide and water and convert them into sugar, starch, and other compounds, together with oxygen which they then give out to the air. Animals, by eating plants, absorb these compounds, which react with oxygen (breathed in from the air) and form carbon dioxide and water; these are then returned to the air when an animal breathes out, and so they are once again available to plants. You can see that there is a constant exchange of these compounds between plants and animals, and you will learn more about it in later science. The most important of the compounds taking part in this sort of merry-go-round is carbon dioxide, because the element carbon, by its special properties, is the basis of plant and animal (organic) compounds. So the process is called the *carbon cycle*.

See if you can write a word equation to show the chemical reactions of the carbon cycle as depicted in Figure 8.9.

In the chemical industry carbon dioxide is produced in various ways: by burning coke (impure carbon), by heating limestone (calcium carbonate), and in beer and wine-making (as a by-product).

Properties of carbon dioxide

The gas has no colour or smell, is denser than air, and is slightly soluble in water.

Its chemical properties are:

(a) It does not burn.

(b) It does not support the combustion of ordinary materials, e.g. fuels, and is therefore used to extinguish fires; being denser than air it tends to remain longer on the fire than would a less dense gas, thereby restricting the supply of air to the flames. However, reactive metals, e.g. magnesium, burn in the gas, by decomposing it and combining with the oxygen (see Chapter 15).

(c) When carbon dioxide is dissolved in water a small part of it reacts with the water to form carbonic acid (from which the salts, carbonates and hydrogen carbonates are derived). As there is so little acid in the water, the solution has the properties of a weak acid.

(d) When carbon dioxide reacts with lime water (calcium hydroxide solution), the insoluble white compound, calcium carbonate, is formed. This is the cause of the milkiness in the lime water test for carbon dioxide. When an insoluble substance is formed in this way in a solution by chemical reaction, it is called a *precipitate*.

Oxygen

Oxygen is essential to both plant and animal life and makes up about one fifth of the air. Things cannot burn without it, and it is used extensively for producing the high temperature oxy-acetylene flame to weld and cut metals. Oxygen has to be supplied to divers and astronauts, and sometimes to patients in hospitals. Mountaineers need it at great heights where there is less air. You already know how oxygen is made industrially. In the laboratory it can be made by heating various oxygen compounds such as mercury oxide, red lead oxide, potassium permanganate, potassium chlorate. The last two give the most oxygen. Potassium permanganate has the disadvantage of making brown stains on apparatus and hands, and potassium chlorate gives its oxygen very slowly. However, if the compound manganese dioxide is mixed with the chlorate, oxygen is evolved rapidly. The manganese dioxide (a black substance) remains unchanged after the experiment and it is the potassium chlorate (white) which produces the oxygen. A substance such as manganese dioxide which changes the rate of a chemical reaction but remains unchanged itself after the reaction is called a *catalyst*. This method of making oxygen should not be used unless the purity of both the potassium chlorate and manganese dioxide is first carefully checked, since explosions have been caused by contamination. The simplest and usual method of preparing the gas is the one described in Experiment 8.6, using hydrogen peroxide and manganese dioxide as a catalyst. The hydrogen peroxide decomposes into water and oxygen.

Burning — combination of substances with oxygen

In Experiment 8.6 you found that substances burn much better in pure oxygen than in air, which contains only about 20% of the gas.

Carbon dioxide being used to put out a fire

Pure oxygen being used during an operation

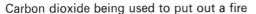

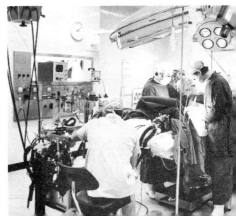

The property of oxygen to re-light a glowing splint is used as an identification test for the gas.

The *sulphur* burned to form a gas called sulphur dioxide. It is an acidic oxide because it reacts with water to form an acid (sulphurous acid). This was the acid which caused the blue litmus to turn red.

The *carbon* formed the gas carbon dioxide, some of which reacted with the water to form carbonic acid. This turned the blue litmus red.

Iron formed the hard, insoluble iron oxide when it burned, and *magnesium*, a reactive metal, combined with the oxygen so vigorously that much heat and light were evolved. The product was magnesium oxide which is slightly soluble in water. It reacts with the water, forming the alkali magnesium hydroxide, which turned the phenolphthalein pink.

In Experiment 8.7, the metals *sodium, calcium,* and *zinc* burned with coloured flames: golden yellow, brick red, and bluish-green respectively. Sodium and calcium are more reactive (chemically strong) than magnesium; their oxides are more soluble than magnesium oxide and the alkalis—formed by these oxides reacting with water—are stronger than magnesium hydroxide. Zinc formed white zinc oxide when it burned, and wisps of this light powder float about in the air (hence the old popular name 'philosopher's wool'). The oxide becomes temporarily yellow when heated, and is insoluble in water.

Oxygen is evidently a reactive element which combines with both metals and non-metals, forming oxides. When oxygen is added to a substance chemically like this the process is called *oxidation*. There are different classes of oxides, and four of them are particularly important:

(a) A *non-metal* combines with oxygen to form an oxide which, when it reacts with water, makes an acid. Such oxides are therefore called *acidic oxides*.

(b) A *reactive metal* (e.g. calcium) forms an oxide which reacts with water to form an alkali. This type of oxide is therefore termed an *alkaline oxide*.

(c) The majority of metals such as iron and zinc form oxides which are insoluble in water and do not form alkalis.

(d) Some non-metals can also form *neutral oxides*, such as hydrogen oxide (water) and carbon monoxide. These do not form acids with water.

You will learn more about oxides and their classification later.

Hydrogen

Hydrogen is not one of the gases of the air, but is contained in many compounds such as water, acids, sugar, and starch. It is a highly flammable gas and for this reason it has been replaced by another light gas, helium, for use in airships. Considerable quantities of hydrogen are used for making ammonia gas (page 73). It is obtained from water by various methods, one of which you will learn about in Chapter 12.

Properties of hydrogen

In your experiments you will have noted the obvious physical properties of the gas: it has no colour or smell, has a very low density, and is insoluble. Several of its chemical properties were also shown:

(a) It forms an explosive mixture with air, even more so with oxygen. Any flame or glowing substance should therefore be kept well away from it, unless some of the gas is being purposely exploded as in Experiment 8.8. If hydrogen and oxygen are mixed in their correct combining volumes (two of hydrogen to one of oxygen) the explosion has considerable force.

(b) Hydrogen does not support combustion. (Why?)

(c) The gas burns with an almost invisible flame to form water. This is also formed when hydrogen explodes (burns rapidly). You should now be able to explain the wax taper experiment, if you have not already done so.

(d) Because hydrogen combines easily with oxygen it is sometimes used to remove oxygen from a compound. The compound is then said to be *reduced* by the hydrogen which is called the *reducing agent*, and the process is called *reduction*. A good example is the reduction of copper oxide by hydrogen, i.e. the removal of its oxygen, in order to produce copper (Demonstration 8.10).

Questions

1. How do you know that air is mainly a mixture of oxygen and nitrogen and not a compound of these elements? What other gases are there in clean air?
2. Describe how you could demonstrate by two simple experiments that air contains (a) carbon dioxide and (b) water vapour.
3. In an experiment to show the approximate proportion of

oxygen in the air, iron filings were stuck to the bottom of a measuring cylinder which was then inverted in a bowl of water. What was the function of the iron? What was in the cylinder, apart from water, at the end of the experiment?

4. What kind of oxides do you think would be formed if (a) phosphorus (a non-metal) and (b) potassium (a light reactive metal) were burned in oxygen? Would the oxides react with water? If your answer is 'yes', name the product formed; if 'no', explain your answer.

5. State the approximate proportions of the gases in clean dry air, and explain briefly how air gets polluted.

6. Describe briefly how the following gases are made and identified in the laboratory: (a) carbon dioxide, (b) hydrogen, and (c) oxygen.

7. Mention two important uses of each of the gases in question 6, explaining how the use depends on a property of the gas.

8. In an experiment to find the density of a gas under laboratory conditions of temperature and pressure, a pupil collected 75 cm³ of the gas by heating 7.65 g of a chemical. The weight of the chemical after heating and cooling was 7.41 g. What was the density of the gas?

9. What are the noble gases in the air? Describe the uses of two of them.

10. Which of the following elements, when burnt in oxygen, (a) gain in mass or (b) decrease in mass: sulphur, magnesium, carbon, iron? Explain your answers.

11. Give examples of, and explain what is meant by, (a) an alkaline oxide, (b) an acidic oxide, and (c) a neutral oxide.

12. Give two physical and two chemical properties of (a) hydrogen, (b) oxygen, and (c) carbon dioxide.

13. Describe briefly how to find by experiment the proportion of oxygen in the air.

14. Draw a labelled diagram to show the apparatus for making oxygen and collecting gas jars of it.

15. Name the products formed when the following elements burn in oxygen: (a) sulphur, (b) zinc, and (c) calcium. Describe what you see in each case. What happens when the products are mixed with water?

16. How would you show that (a) hydrogen does not support combustion? (b) hydrogen, when not mixed with air or oxygen, burns quietly?

17. Explain, by brief reference to an experiment, what is meant by describing hydrogen as a reducing agent.

18. Draw a labelled diagram of the apparatus for demonstrating that when hydrogen burns, it forms water. Assume that the hydrogen comes from a cylinder of the gas.

19. Explain briefly what is meant by the *carbon cycle*.

9 Smashed to Atoms

Discussion and Investigation

Figure 9.1

Atoms

Today, in an increasingly scientific world, everybody knows that all substances (matter) consist of tiny things called atoms. We hear and perhaps read about atomic bombs, atom smashers, and atomic power, and see television programmes about them. Obviously atoms are powerful and important, dangerous too. You cannot learn chemistry properly or understand about chemicals and the way they work unless you know something about atoms. But it is not easy, at first, to understand how a hard, smooth object, such as a block of metal or a piece of glass, can be made of millions of little separate parts, *particles* as they are called. Particle is a useful word in science because it can refer to atoms, molecules (see page 86) and other tiny bits of matter.

Do you find it easier to accept this idea of particles in the case of powdery things like earth, sand, flour, etc.? Think of other examples of powders, and then try to work out whether any of them have come from hard things like rocks and metals; sawdust from wood and iron filings from pieces of iron are examples. What about sand and flour?

The first people to think up the idea of particles were the Greeks, who reasoned that it must be possible to cut a piece of something, e.g. iron, into smaller and smaller pieces until a minute piece is obtained which cannot be cut up any more. Their word for this tiny piece, 'atom', is Greek for 'uncuttable' or indivisible. So they considered that if we could magnify things thousands of times we should see that they really consist of huge numbers of these little particles.

Have you ever looked closely at a smooth tarmac road? You will find that it consists of masses of little pieces of stone rolled closely together with tar, rather like a jigsaw puzzle. Can you think of other examples of things which appear to be all one piece but are really made of smaller parts? Perhaps you have thought about this when looking through a microscope. Magnified in this way, the apparently smooth, straight edge of a used razor blade can look like a jagged line of miniature mountain peaks, and snow particles look rather like flowers with six petals. Have you ever seen anything under a microscope which did not look entirely different from its ordinary appearance? Discuss some of the things you have seen magnified in this way. Do water and other liquids look different through a microscope?

A magnified snowflake

If water is made of particles they must be too small to be seen with a microscope. Do you think that liquids also consist of particles, or is their matter just one complete thing? Compare solids and liquids. Solids can be cut up, but can liquids? Can solids be poured from a jug like liquids? Can solids change their shape as liquids do? A lump of rock can be ground to a powder such as sand (from quartz), and sand can be poured like a liquid. Can it change its shape?

Experiment 9.1

Pour some marbles from one container into another. Try small beads or rice. They behave more like a liquid than marbles. Now try very fine, dry sand, or caster sugar, or dry salt. These are even better. From a distance the sugar (or salt) looks like a white liquid being poured. Listen to the sound when each substance is being poured. You can easily hear the separate sounds of the marbles, but a fine powder sounds almost like a liquid. Does this experiment suggest to you that liquids might also consist of tiny particles?

Experiment 9.2

Fill a glass with water right to the brim so that anything else added would make the water overflow. Now add *slowly* a teaspoon of dry, free-flowing salt. Does the water overflow? Add more salt. What has happened to the little salt particles? Now do the experiment again, but instead of water, fill the. glass with marbles. Are there spaces for the salt, even when the marbles come right to the top of the glass? Now can you explain why the water did not overflow?

Experiment 9.3

Place a very small grain of potassium permanganate in a beaker of water. Does it split up into many tiny particles? Stir the water until it has a definite colour. What causes the colour? Now pour the coloured solution into a larger beaker and add more water. Does the colour, though paler, remain?

This experiment may give you some idea about the very small size of particles.

Experiment 9.4. To show how a powdered solid resembles a liquid

Clamp a Buchner filter funnel without filter paper to a retort stand, and connect it by rubber or plastic tubing to a small pressure pump. Fill the funnel with fine, dry sand, just over half full. Blow air through the funnel. The sand moves and ripples like a liquid, and small objects move about when floated on it. If they are pushed under the sand they rise to the surface again.

It is much easier to think of a gas as consisting of particles. Do you know of any experiment which might suggest this? Watch steam coiling upwards in various directions and splitting up easily into separate bits. We cannot see air, but can feel it by waving our hands about. If you watch a tree swaying in the wind you will notice that some of the leaves move while others do not. Or sometimes parts of the wind curve round, blowing things helter skelter in all directions.

By discussion and thinking about the subject you are probably satisfied now that matter does indeed consist of millions of minute particles, even though they cannot be seen. There is one more problem for you. You have learnt about elements and compounds. When one element combines with another (for example, magnesium with oxygen), and a new substance is formed (magnesium oxide in our example) do you think that particles play a part?

Further Information

Particles
When scientists first described solids, liquids, and gases as consisting of particles, their ideas were called the 'Kinetic Theory of Matter'. Their investigation and reasoning were along the same lines as ours. One of the first things of practical significance to give them the particle idea was the fact that when water and ethanol are added together they mix so thoroughly that even the tiniest drop of the liquid contains both compounds. This could only be explained if both consisted of particles with spaces between them. Modern experiments with complicated apparatus have proved the guesses of these early scientists to be true. On your physics course you will learn about the way particles move about in liquids and gases, and how pressure and other phenomena are explained by this movement. In chemistry we are concerned more with the nature of the particles themselves, since this helps us to understand the properties of elements and compounds.

Atoms
The particles of elements are called atoms, and different elements have atoms of different sizes and masses. The atoms of a particular

element are all the same. An atom is the smallest part of an element that can take part in a chemical reaction. When magnesium, for example, combines with oxygen to form magnesium oxide, each of the millions of atoms in the piece of magnesium combines with an oxygen atom.

Atoms are far too small to be seen with a microscope. In fact they cannot be seen at all. But scientists have learnt a great deal about them by very skilful and advanced techniques. In one experiment the paths of atoms can be studied as they move about in a container called a 'cloud chamber'. It is rather like watching a vapour trail high up in the sky when you cannot see the aircraft which is making it. As to size, you will get some idea if you consider a speck of iron (from iron filings) about the size of the average full stop. This speck contains over two million iron atoms. The weights of atoms are just as small. The lightest (and smallest) atom is the hydrogen atom. About 10^{18} hydrogen atoms weigh one gram, which is the smallest brass weight in the laboratory.

Inside atoms there are even smaller and lighter particles. These are called *neutrons*, *protons*, and *electrons*. The neutrons and protons are in the centre or *nucleus* of the atom, and the electrons buzz about just outside the nucleus, rather like a small cloud of steam around a speck of dust floating in the air. It is these electrons which enable atoms to combine together to make compounds, so electrons are very important in chemistry.

In electricity you learn about the positive end and the negative end of a battery, and you will discover that an object can be given a

The vapour trail of a large aircraft

positive charge or a negative charge of electricity, and that two objects having different charges attract each other, like the north pole of one magnet attracting the south pole of another. The protons in an atom are positively charged, and the electrons are negative. So the electrons are attracted to the nucleus (containing the protons) and do not spin away. The neutrons have no charge. The number of protons in an atom of an element is always the same as the number of electrons, so the opposite electrical charges balance out and the atom itself is neutral. This is a good thing, otherwise we should get a shock if we could touch something consisting of millions of charged atoms! Some compounds do

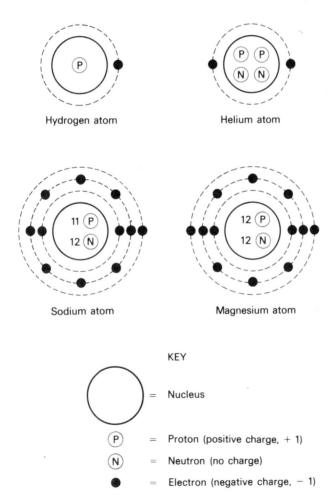

Figure 9.2 Diagrammatic representation of atoms. Note carefully that these two-dimensional diagrams only show the numbers of protons and neutrons in the nucleus and the approximate arrangement of electrons at varying distances from the nucleus. They are *not* pictures of atoms.

contain charged atoms, called *ions*, but there are always equal numbers of positive and negative charges, so again there is no danger of getting an electric shock. Charged atoms (ions) get their charges by either losing electrons and becoming positively charged or by gaining electrons and becoming negatively charged. You will learn about this when you study electricity in more detail and find out how it is concerned with chemistry.

Molecules

The atoms of many elements group themselves together to form *molecules*. The molecules of most gases which are elements, e.g. oxygen, hydrogen, nitrogen, etc., consist of pairs of atoms. Some solid elements also consist of molecules. Sulphur molecules have eight atoms for example, and there are four atoms in the phosphorus molecule. The atoms in the molecules of a particular element are of course identical. In compounds, the atoms are different, being those of the elements in the compound. The water molecule contains two hydrogen atoms and one oxygen atom, and the carbon dioxide molecule contains one carbon atom and two oxygen atoms. Obviously, the smallest particle that a compound can have is its molecule. If the water molecule were split up it would become hydrogen atoms and oxygen atoms.

Questions

1. You are explaining to a friend, who knows nothing of the subject, that solids, liquids, and gases consist of tiny particles. Write a short account of what you would say to him and the experiments you would mention.
2. Explain why it is possible to add salt to a glass of water filled to the brim without any water spilling from the glass.
3. What is an atom? Describe the main particles which are inside atoms.
4. How would you describe the very small size of atoms?
5. The atoms of some elements group themselves together. What are these groups of atoms called?
6. How many atoms are there in groups of atoms in the case of the following elements: (a) oxygen, (b) sulphur, (c) nitrogen, and (d) phosphorus?
7. What is meant by the term *ion*?
8. What kind of particle is formed when an atom of carbon combines chemically with two atoms of oxygen?
9. What are the atoms in a molecule of water?

10 Good and Bad Behaviour

Discussion and Investigation

Figure 10.1

You have now carried out many experiments in which elements and compounds change into new substances. Sometimes the change is caused by heating the substance, and sometimes two or more substances react together, as in the neutralization reaction between an acid and an alkali.

Most chemicals change fairly gently, without any kind of fuss, but sparks sometimes fly, as in explosions. Too much heat can make gases evolve too quickly and dangerously, so you should always take the greatest care, especially when heating substances in test tubes. It is better not to boil a liquid in a test tube because it can suddenly shoot out and scald you or, worse still, somebody else.

Read through the second half of Chapter 4 and Experiment 8.6, and then discuss the results of your experiments. You will notice that the substances which changed permanently when heated did so in two ways. Can you describe these two particular types of chemical behaviour? Give examples from your experiments. A third type of chemical reaction occurred when you made hydrogen from zinc and dilute sulphuric acid. How would you describe it? It may help you to write a word equation. If you are learning about chemical equations and have studied Chapter 16 you will be able to write the proper kind, using symbols and formulae. You probably explained the reaction correctly, in which case you will be able to guess what will happen in the following experiment.

Experiment 10.1

Place a piece of zinc in a small beaker of copper sulphate solution. After about five minutes take it out. What do you see? What is the other substance which has formed, and where has it come from? Write an equation to explain the reaction. If several pieces of zinc were left in the beaker of solution for some time, do you think there would be any change in the colour of the solution?

You made copper oxide, a black compound, by two quite different methods. In one, you heated copper carbonate, and in the other you heated the metal, copper. Is the copper oxide in each case exactly the same substance? Or do you think that it differs, depending on how it is made? Explain your answer carefully.

In Experiment 4.6 you found that blue copper sulphate lost mass when heated, due to the water driven off. Do you think that the water accounts for *all* the lost mass? If this water had been more carefully collected and weighed, would you have found the following statement to be true or untrue?

weight of water + weight of anhydrous copper sulphate
= weight of blue copper sulphate

Was any matter lost or destroyed during your experiment? Is matter destroyed during any experiment?

Have you thought about the causes of chemical changes? What really happens when a compound decomposes or an element combines with another? What happens to the atoms of the substances? And why is heat so often necessary?

Can you suggest how chemical changes can be made to take place more quickly? If a substance—wood for, example—is chopped up into small pieces does it burn more rapidly? Can this method of speeding up a chemical change be applied to any of the experiments you have already done? Do you think that a chemical change happens more rapidly when a substance is heated more strongly? Do some chemical reactions give out heat once they are started?

Explanation and Further Information

You will have realized during your experiments that a *chemical change* is very different from simple changes like water turning into steam and salt dissolving in water. Such changes are called *physical changes* and can easily be reversed or changed back. Steam can be turned into water again merely by cooling it, and salt can be obtained from salt water by simple evaporation. No new substances have been formed. Chemical changes are obviously much bigger, more drastic, and usually they are not easily reversed. The burning of magnesium is a good example. The metal burns with a dazzling light, and an entirely new substance, the compound magnesium oxide, is formed. The properties of the oxide are quite different from those of its elements, a metal and a gas. And it is difficult to reverse the change, to get the magnesium and oxygen out of the oxide.

A chemical change is caused by a chemical reacting to heat (or other forms of energy) or with another chemical. If the chemical is a compound it may react to heat, light, etc. by decomposing, as you have discovered. In a *decomposition reaction* a compound splits up into its elements or into simpler compounds:

e.g.   mercury oxide + heat → mercury + oxygen

copper carbonate + heat → copper oxide + carbon
dioxide

Or a compound may react with another chemical, combining with it to form a more complex compound:

e.g.   water + sulphur dioxide → sulphurous acid
or H_2O + SO_2 → H_2SO_3

If the substance is an element it can only change by combining with another substance and this reaction is often started or speeded up by heat:

e.g.   copper + oxygen → copper oxide

In these last two examples, the type of chemical reaction is called *combination*. Now check your answers to the first question on page 87.

In a third type of chemical reaction one element replaces another in a compound. This happened in Experiment 10.1:

zinc + copper sulphate → copper + zinc sulphate
Zn + $CuSO_4$ → Cu + $ZnSO_4$

Other examples are:

$$\text{zinc} + \text{sulphuric acid} \rightarrow \text{hydrogen} + \text{zinc sulphate}$$
$$Zn + \quad H_2SO_4 \quad \rightarrow \quad H_2 \quad + \quad ZnSO_4$$

In these *replacement* reactions zinc is chemically stronger (more reactive) than both copper and hydrogen, so it is able to 'push' them out of the compounds, forming zinc sulphate in each case.

When a chemical reaction can be reversed, as in the example concerning copper sulphate shown by the word equation on page 27 (Chapter 4), it is called a *reversible reaction*. As you learn more chemistry you will find out about other types of chemical reactions.

Elements and compounds do not vary in any way by being made by different types of chemical reactions. If you revise elements and compounds in Chapter 5 you will realize that this must be so. So the black copper oxide you made from copper carbonate (Chapter 4) was exactly the same as the black copper oxide formed when you heated copper.

Nothing is lost or destroyed in a chemical reaction, nor can any extra matter be made. The mass of the *reactants* (substances reacting) is always the same as that of the *products* (substances formed). This is summarized in the *Law of Conservation of Matter* which simply states that *in a chemical reaction matter cannot be created or destroyed*.

The word 'law' usually makes one think of police and law courts, but in science it is used to describe something which always happens without any exception, like liquids turning into solids when their temperature is lowered sufficiently. If scientists are not sure and think that what seems to be a law has not yet been fully proved, they call it a theory. You will meet with other chemical laws as you continue to study chemistry. Sometimes it seems that matter has somehow been lost in a reaction, as when a candle burns down to almost nothing. Candle wax is a mixture of compounds of hydrogen and carbon, and when it burns these elements combine with oxygen to form water and carbon dioxide respectively. The total weight of the candle before burning and the oxygen with which it combines is exactly equal to the combined weights of the water, carbon dioxide, and small residue left after burning.

You will realize now that the statement given on page 88 is true. If copper is heated in a closed test tube, some of it combines with the oxygen of the air in the test tube, forming a coating of black copper oxide. If the test tube is weighed before and after the experiment it will be found that there is no change in mass. Nothing has left the test tube or gone into it. Some of the copper atoms have simply joined up with some oxygen atoms.

What really happens, then, in chemical reactions is that the atoms of the reactants rearrange themselves so as to form the products. When hydrogen atoms react with copper oxide, for example, they and the oxygen atoms are rearranged to form water, and the copper atoms, previously combined with oxygen atoms, become free as copper metal. If the metal, iron, is placed in sulphuric acid (hydrogen sulphate), another compound, iron sulphate, and hydrogen are formed. The sulphate group of atoms (sulphur and oxygen atoms) does not change. Which atoms do?

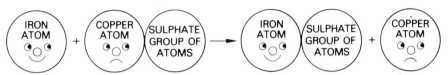

Figure 10.2 A chemical reaction—rearrangement of atoms

When atoms rearrange themselves in a chemical reaction they either give out energy or take it in. In most cases the energy is in the form of heat. When heat is given out, the reaction is called *exothermic*, an example being combustion or burning. You have burnt several substances in air or pure oxygen and will have noted that heat is evolved, as indicated by the flames or glowing of the substance:

e.g.   Sulphur + oxygen → sulphur dioxide + heat
           S   + O_2 → SO_2 + heat

The majority of chemical reactions are exothermic.

An *endothermic reaction* is one which absorbs heat. This means that the reactants must be heated all the time to make the reaction continue. You will remember having to heat (add heat to) green copper carbonate to make it decompose (Experiment 4.2):

Copper carbonate + heat → copper oxide + carbon dioxide
     $CaCO_3$ + heat →   CuO   + CO_2

Make a list of chemical reactions in experiments you have done or seen, indicating whether they are exothermic or endothermic. The following are examples:

Exothermic	*Endothermic*
Burning of magnesium	Decomposition of copper carbonate

You can probably recall a reversible reaction, in one of your experiments, which is exothermic in one direction and endothermic in another.

It is important to note that both exothermic and endothermic reactions go faster when the temperature is raised.

It will now be clear to you why some substances have to be heated all the time to cause a chemical change, and why others, like magnesium, give out heat when they change into new substances. But you may wonder about the necessity to heat the magnesium a little before it burns of its own accord. Why is a match needed to light a fire? For a complete explanation you will have to wait until you have learned more chemistry. The simple answer is that some substances need what is called activation energy to start them reacting. It is rather like having to push a toboggan before it will slide down a slope.

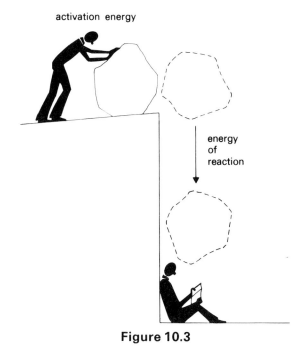

activation energy

energy
of
reaction

Figure 10.3

There are other ways, in addition to increasing the temperature, of making a chemical reaction go more quickly. One is to increase the surface area of the reactants by breaking them into powders. Iron filings, for example, react with an acid much better than an iron nail. If the particles are even smaller, as they are in solutions, the reaction is quicker still. You will have noticed how rapidly an acid reacts with a solution of an alkali.

Questions

1. A physical change is a simple process which can easily be reversed, but a chemical change is more complicated. Explain this statement, with examples.
2. What type of chemical reaction is represented by each of the following examples?
   (a) Water + carbon dioxide → carbonic acid
       H_2O CO_2 H_2CO_3
   (b) Hydrochloric + sodium → sodium + water
       acid   hydroxide   chloride H_2O
       HCl   NaOH   NaCl
   (c) Calcium carbonate → calcium oxide + carbon dioxide
       $CaCO_3$ CaO CO_2
   (d) Sulphur + oxygen → sulphur dioxide
       S O_2 SO_2
   (e) Magnesium + zinc sulphate → magnesium sulphate + zinc
       Mg $ZnSO_4$ $MgSO_4$ Zn
   (f) Mercury oxide → mercury + oxygen
       $2HgO$ $2Hg$ O_2
3. Write a word equation to show a reversible reaction.
4. A student made some magnesium oxide by burning magnesium, but was criticized by his friend who said that a better way was to heat magnesium carbonate because the magnesium oxide formed this way was more soluble and contained more oxygen. Do you think his friend was right?
5. What is the Law of Conservation of Matter? Explain with an example how it applies to chemistry.
6. Explain what is meant, giving examples, by (a) an exothermic reaction, (b) an endothermic reaction.
7. When the elements carbon and oxygen combine chemically, they give out a great deal of heat, yet heat has to be put into the carbon before it will burn. Explain this.
8. Describe two ways of making a chemical reaction go faster.
9. A boy read his brother's 'O' level chemistry book about ammonia and found that it is made by the combination of nitrogen and hydrogen. He also read that if the temperature is too high, some of the ammonia splits up into nitrogen and hydrogen again. He said to his brother: 'This chemical reaction, the making of ammonia from its elements, must be one which is *not* made to go faster by raising the temperature, so my book must be wrong.' What do you think? Can you offer any explanation as to why both books might be right?

11 Fire and Flames

Discussion and Investigation

Fire was a necessity to mankind even in prehistoric times. It provided warmth, and protection against animals. The fierce life-like flames of a fire were as much a mystery then as they were later to the Greek philosophers, who thought that there were only four elements: earth, air, fire, and water. Even Lavoisier, in the eighteenth century, classified heat as an element.

Some people are just as curious today about fire. What do you think about it? Why do flames have different colours? You have

The Olympic flame at the 1976 Olympic Games in Montreal

burnt substances in air and oxygen; did these experiments give you any ideas about what causes fire and what a flame consists of? See if you can answer these questions:

1. Why did you have to heat sulphur for a little while before it caught alight?
2. When you are lighting a candle, how do you explain that short wait before the flame comes?
3. Why does a candle need a wick?
4. If you concentrate the sun's rays on a piece of paper with a magnifying glass (lens), the paper begins to char and smoke. Why is there a delay before you can get a flame?
5. Do substances always have to be heated to burn?
6. Do some have to be heated more than others?
7. Do some substances refuse to burn, however much they are heated?

The elements that you burnt or heated strongly formed compounds called *oxides* by chemically combining with oxygen. Do you think that compounds and mixtures form oxides like this when they burn? Wood, oil, paper, and coal burn well and give out plenty of heat. Is their burning the same kind of chemical reaction as magnesium or sulphur burning?

After thinking about the questions asked above you probably concluded correctly that if a substance is heated sufficiently it sometimes forms enough gas to make it burst into flames, and that flames therefore consist of very hot gases. Assuming that all normal types of burning are due to substances combining with oxygen and producing oxides which are often gases, like sulphur dioxide, can you work out a method by which a substance such as wood could be burned and the gases trapped and identified? You have already learned how to identify several substances.

Experiment 11.1. **To study the burning of a candle**

Stick a small candle on the glass cover of a gas jar. Light the candle and observe carefully the different parts of the flame and how it burns. Write down all your observations. Now place a clean, *dry* gas jar over the flame. The candle soon goes out, but what else do you see? Holding the cover and candle firmly to the gas jar, turn the jar over and place it right way up on the bench. Add a little clear lime water to a test tube, remove the cover and candle, and pour the lime water into the gas jar. Quickly place another cover over the jar, and shake. What happens to the lime water? Now complete your written observations.

Experiment 11.2

Repeat Experiment 11.1, but burn a piece of wood splint instead of the candle. Stick it on the glass cover with candle wax so that it stands upright, and bend the top down, partly breaking it, to make it burn more easily.

If you observed the candle flame carefully, you probably noticed certain distinct parts or regions of the flame. In the next experiment you can make some interesting discoveries about the regions of the Bunsen flame.

Experiment 11.3. To study the regions of a Bunsen flame

1. Light the burner with the air hole closed. You should always light it this way otherwise the flame sometimes burns at the jet at the bottom. Examine the regions of the flame, and then hold a white evaporating basin or crucible in the yellow flame for a few moments. What happens to the white surface? Now repeat this when the air hole is open and the flame is almost non-luminous. Is the white surface affected in the same way?

2. Hold a piece of light-coloured wood over the non-luminous flame, about 4 cm above the burner, for a few moments. Describe the markings that have been made on the wood.

3. Stick a pin through a live match, about a centimetre from the head, and place the match in a Bunsen burner so that it rests on the pin and the head is exactly in the centre of the burner. Close the air hole and light the burner. What happens?

4. Adjust a Bunsen flame to give it the maximum amount of inner blue cone. Hold a piece of soft glass tubing: (a) in the top of the flame; (b) in the blue cone; and (c) just at the tip of the cone. You can tell which is the hottest part by the approximate time it takes for the glass to glow and turn the flame partly golden.

Explanation and Further Information

Combustion

The scientific name for burning is *combustion*, which is a chemical reaction, normally involving combination with oxygen, in which heat and usually light are given out. When a solid or liquid burns with a flame it is because gases have been formed. The chemical reaction evolves so much heat that these gases glow and emit light. The colour of the light depends on the substance burning (see Experiment 8.7). The yellow part of the flames of burning wood, candle wax, gas, and other ordinary substances is caused by glowing particles of carbon which have been released from carbon compounds and are mixed with the burning gases. Some substances, such as carbon, burn without a flame, as you found in Experiment 8.6. Some have to be heated more than others to make them burn, and some, like asbestos, do not burn.

Fuels

Substances which are burnt for the purpose of producing heat or light are called *fuels*. Common fuels are gases, liquid petroleum fuels, oil, wood, and coke (impure carbon). Coal is also used, but has the serious disadvantage of polluting the atmosphere with chemicals which are much too valuable to be lost in this way. In the chemical industry the coal is heated so that these compounds are carefully separated and the residue is smokeless fuel or coke.

Most fuels consist mainly of compounds containing carbon and hydrogen. So when they burn, the two elements are oxidized to carbon dioxide and water, as you found with the candle and wood. Some fuels contain small quantities of other substances which are released by the heat and so we find smoke (minute solid particles), tar, and traces of other gases among the carbon dioxide and water (main combustion products).

The Bunsen burner

When you watched the candle burning you must have noticed black fumes (soot) rising from the flame. Most fuels burn with sooty yellow flames unless special burners are used. Bunsen, a German

A flame-proof asbestos suit in use

chemist, invented the famous Bunsen burner in 1855 not only to give a clean, non-sooty flame but to make the flame hotter. He did this by arranging that extra air mixed with the gas before it reached the flame. You have already learned how to use the Bunsen burner. The adjustable collar at the base of the burner opens or closes a hole through which air can enter. If you look through the hole you will see the small jet whereby a fast stream of gas is made to travel up the burner. The speed of this stream draws in the extra air when the hole is open. Without extra air mixed with the gas, the carbon particles in the middle of the flame cannot be oxidized by contact with the air outside, hence the sooty flame. Can you explain why such a flame is cooler than a blue flame?

Regions of a flame

In Experiment 11.3 you found that the yellow part of the flame produced carbon (in the form of soot), but that this was entirely absent when the flame was blue. A dark, scorched ring was formed on the piece of wood, suggesting that the interior of the flame was not as hot as the outside. This was confirmed when the match inside the flame did not ignite for some time. Where was the hottest part of the flame? Check your notes to see how much you discovered for yourself, and then read the next paragraph for more detail, studying Figure 11.1 carefully.

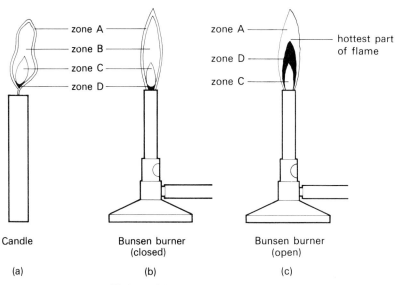

Figure 11.1 Regions of a flame

In the diagrams, you will see that the yellow flames of the candle and closed Bunsen burner are similar, except that the Bunsen flame has a regular shape, not unduly affected by drafts, due to the flow of gases. But the blue flame of the open burner is very different. Zone A is a faintly luminous fringe in (a) and (b), but occupies most of the flame in (c). In this region combustion is complete, i.e. all the carbon and hydrogen are oxidized to carbon dioxide and water by combining with oxygen from the air. Zone B is the yellow flame, absent in (c), in which there is excess carbon. The tiny particles of carbon make the flame luminous by glowing, and some of the carbon escapes from the flame as soot. Zone C is the hollow centre of the flame in which there are unburnt gases. This region is therefore comparatively cool. Zone D is a light blue region in which carbon monoxide is burning. In the open Bunsen burner the top of this blue zone is the hottest part of the flame and can reach 1600 °C.

You will now understand that when extra air is mixed with the gas, there is enough oxygen to oxidize all the carbon, so more heat is produced and no soot forms. For normal purposes, the air hole should be half open and the gas not turned on fully. This gives a tip of yellow to the flame which can then be easily seen.

Rates of combustion

In Experiment 8.6 you found that substances burn more quickly in pure oxygen than in air (since air contains only 21% of oxygen). Evidently the rate of combustion can vary. When the amount of oxygen is just right for a particular substance, the combustion can be so rapid that an explosion occurs. If hydrogen, for example, is mixed with oxygen in the proportion of two volumes of hydrogen

An oxy-acetylene flame used for cutting scrap metal

to one of oxygen, and the mixture is ignited, a loud explosion results. But you caused a smaller, less rapid explosion when you ignited a mixture of hydrogen and air (Experiment 8.8); and you have seen hydrogen just burning slowly and quietly. In a petrol engine, the proportions of petrol gas and air have to be very carefully controlled, so that when the electric spark occurs there is a powerful explosion in each cylinder which causes the engine to turn. If the mixture is wrong, the engine does not start, or it runs badly. It is the same in a compression ignition (diesel) engine, except that there is no sparking device because the gases are made so hot (up to 900 °C) by compression that the fuel ignites by itself.

Combination with oxygen can be a very slow process, as you found when heating copper. Another example is when phosphorus smokes on exposure to the air. Slow oxidation like this is sometimes called 'slow combustion'. When the substance becomes hotter it may suddenly burst into flame and combustion at a normal rate then occurs. This can happen to a heap of dust or rubbish when temperature and other conditions are right.

Oxidation and reduction

Combustion is not the only chemical reaction involving combination with oxygen. In the combustion of carbon the addition of oxygen (*oxidation*) produces carbon dioxide, but carbon can also be oxidized to carbon dioxide by reacting with concentrated nitric acid. In the reaction between hydrogen and copper oxide you learned about *reduction* as the *removal* of oxygen from a compound by a *reducing agent* such as hydrogen (Chapter 8). If you think about it, you will realize that in this same reaction the hydrogen is oxidized to water. The oxygen comes from the copper oxide which is therefore called the *oxidizing agent*. In fact, oxidation and reduction always go together. You will understand this more easily if you study the following equations:

$$\underset{\underset{\text{reduced}}{\longleftarrow}}{\overset{\overset{\text{oxidized}}{\longrightarrow}}{H_2 + CuO \longrightarrow Cu + H_2O}}$$

The hydrogen (reducing agent) reduces the copper oxide to copper by removal of oxygen, and the copper oxide (oxidizing agent) oxidizes the hydrogen to water by the addition of oxygen.

$$\text{oxidized}$$

$$\text{Zn + FeO} \rightarrow \text{Fe + ZnO}$$

$$\text{reduced}$$

Which are the oxidizing and reducing agents in the above reaction?

Questions

1. How would you briefly explain the nature of fire to someone who, having little scientific knowledge, thinks that it is something like an element which comes out of a substance when it is heated sufficiently? You need not describe the regions of a flame.
2. Describe the regions of the flame of a Bunsen burner when (a) the air hole is closed and (b) the air hole is open.
3. How do chemists define combustion? List three common fuels and describe what happens when one of them burns.
4. Describe briefly a simple experiment by which you could demonstrate that a fuel consists mainly of compounds of hydrogen and carbon.
5. How would you show that the interior of a flame is comparatively cool? Why is it cool? Where is the hottest part of a blue Bunsen flame?
6. Which of the following statements is true? Write corrections for statements you consider to be partly or wholly untrue.
   (a) The purpose of the adjusting collar on a Bunsen burner is to concentrate the gas from the jet so that it burns with a hotter flame.
   (b) The yellow part of a closed Bunsen burner flame is caused by carbon particles.
   (c) Coal is a better fuel than coke.
   (d) When wood burns, water and carbon dioxide are formed.
   (e) Hydrogen cannot burn without oxygen, but if it is strongly heated, without oxygen, it may explode.
7. Two school friends were arguing about car engines. One of them said that the petrol flows through a pipe from the petrol tank into the engine cylinders, and when these are full the electric spark explodes the petrol, some of which can always be seen dripping out of the exhaust pipe at the back of the car. Do you think he was right? What do you think happens? Does petrol come out of the exhaust pipe?
8. Mention (a) a gas which burns; (b) a gas which does not burn; (c) a solid which does not burn; (d) a fuel which produces

carbon dioxide but not water when it burns ; (e) an example of slow combustion.

9. State (a) what has been oxidized ; (b) what has been reduced ; (c) the oxidizing agent ; and (d) the reducing agent in the following reaction :

$$Mg + PbO \rightarrow MgO + Pb$$

12 Electric Shocks for Chemicals

Discussion and Investigation

Figure 12.1

Have you ever experienced an electric shock? You may have given yourself one deliberately when studying the Van de Graaff machine in physics. Or perhaps you have had the very dangerous type of electric shock caused by touching 'live' electric wires carrying a strong current. In either case, you get the shock because the current passes through you to earth. You will probably have already learned

The six million volt Van de Graff generator at Aldermaston

about electricity in physics and will know that metals can conduct an electric current. The only non-metal which can conduct is graphite (a form of carbon), and you may have tested the graphite ('lead') in a pencil to prove this. When these substances conduct a current nothing happens to them; they are exactly the same after the current has passed through them.

It may surprise you to learn that some *compounds* can also conduct a current, but they are changed by the current. In fact, they get such a nasty shock that they are decomposed. Other types of compounds do not allow a current to flow through them. In the following experiment you learn how to test compounds to see whether or not they can conduct an electric current. If they do, you should look for signs of decomposition, i.e. new substances being formed, such as gases.

Experiment 12.1
Apparatus
You need a power pack or six volt battery (or six torch batteries connected together), one six volt bulb, two carbon (graphite) rods (called electrodes), one two-holed cork, a small beaker, and three leads fitted with crocodile clips. Assemble this apparatus as shown in Figure 12.2. The electrodes are fitted into the two-holed cork, in order to space them about half a centimetre apart. Test the circuit by holding the electrodes momentarily on a metal surface, so that they are connected and a current can pass. If the bulb lights up, you will know that the circuit is working properly.

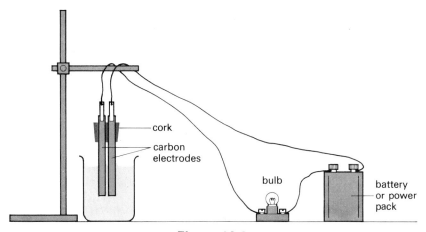

Figure 12.2

Method
Place the compound in the beaker and then put the electrodes
into it. If the bulb lights, the substance is conducting the
current. If a solid does not conduct, make it into a solution and
try it again. When the current is passing through the substance,
observe the electrodes carefully for any signs of chemical
reaction. If there is nothing to be seen, let the current flow for
about five minutes and then remove the electrodes and
examine them. Test the following compounds in the order
shown:
(a) hydrochloric acid, (b) sugar, (c) ethanol (methylated
spirits), (d) copper chloride.
Wash and dry the electrodes and beaker thoroughly before
each substance is tested.

Results
Record the following information:
(a) Whether a liquid conducts electricity.
(b) Whether a solid conducts electricity.
(c) Whether a solid, when made into a solution, conducts
    electricity.
(d) A description of any signs of decomposition at the
    electrodes.

In the last experiment you tested two solids, turning them into
solutions if they did not conduct in the solid state. You could have
heated them, to make them melt into liquids, providing this did not
cause them to decompose. You can try this in the next experiment.

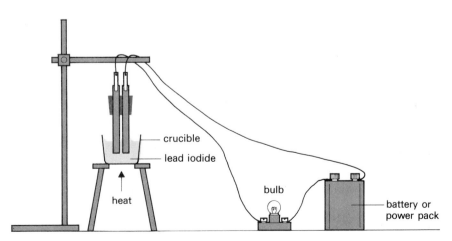

Figure 12.3

Experiment 12.2. **To find out what happens when an electric current is passed through fused (melted) lead iodide**

(For more experienced pupils.)

This experiment should be done in a fume cupboard because **lead fumes are poisonous.**

The apparatus is shown in Figure 12.3. Fill the crucible with lead iodide crystals. Before heating, switch on the current and dip the electrodes into the iodide to see whether current will pass through it as a solid. Then melt the iodide and dip in the electrodes again; does the bulb light? Now disconnect the bulb to give more current. What do you see at the positive electrode? After about five minutes, remove the electrodes and the heat. When the crucible is cool, tap out the lead iodide and examine the bottom of the solid lump. Do you see anything embedded in the iodide? What is it? Remove it and see if it marks paper.

In an apparatus called a *voltameter,* any gases evolved by the decomposition of a compound in solution can be collected and tested. There are various kinds of voltameters, but a simple apparatus consists of a large beaker for holding the solution, two test tubes for collecting the gases, and two plastic coated steel wires for taking the current into the liquid. The wires are bared for 3 or 4 cm at the ends and the non-plastic ends bent back so that they protrude into the inverted test tubes. The beaker is nearly filled with the solution and then both test tubes are completely filled with it, inverted, and placed on to the ends of the wires (electrodes).

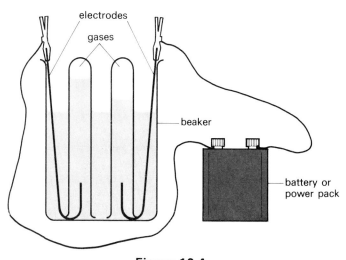

Figure 12.4

To do this successfully a cork or your thumb has to be placed over each test tube while it is being inverted. Water, with a little acid added to it, can be decomposed into hydrogen and oxygen in this way. The apparatus is shown in Figure 12.4.

Demonstration 12.3. **To decompose acidified water by electrolysis**

It is preferable to use either the apparatus described above or, better still, a large glass cylinder containing two burettes and platinum electrodes, as obtainable from laboratory apparatus suppliers. The hydrogen and oxygen can be collected from the burettes in small test tubes and tested in the usual way.

Discuss the results of your experiments and try to think out a simple explanation as to why an electric current can decompose a substance. Remember that electricity is a form of energy, like heat, and read page 89 of Chapter 10 again. When you have come to some conclusion write down your explanation. It need only be in general terms. If you find the problem too difficult perhaps you can explain briefly the difference between a copper wire conducting electricity and an acid doing the same thing.

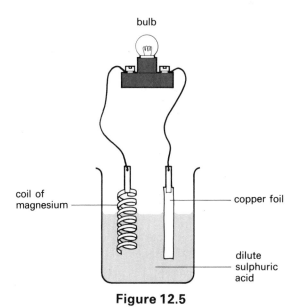

Figure 12.5

If the flow of electrons (i.e. an electric current) through a compound can cause a chemical reaction, perhaps the reverse can happen and a chemical reaction can give out a current of electrons. Here is an experiment for you to do, to see whether this is the case.

Experiment 12.4. **To find out whether a chemical reaction produces electricity**

Connect a bulb to two crocodile leads. To the other ends of the leads attach a coil of magnesium ribbon and a piece of copper foil (Figure 12.5). Now dip the metals into dilute sulphuric acid. What happens?

Explanation and Further Information

Certain compounds, notably acids, alkalis, and salts, can conduct a direct electric current and are called *electrolytes*. When they conduct the current they are decomposed, and this process of decomposition by electricity is called *electrolysis*. In an ordinary decomposition reaction heat is put into the compound and splits it up, so it is not surprising that electrons, another form of energy, can have the same effect if they flow into an electrolyte.

Compounds, like sugar and ethanol, which do not conduct an electric current are called *non-electrolytes*. Do not confuse the terms electrolytes and non-electrolytes with conductors and insulators (non-conductors). A conductor is a substance (usually a metal) which allows a current to pass through it and is unchanged afterwards. An electrolyte is a *compound* which is chemically changed by the passage of electricity through it. An electrical insulator is any material (compound or mixture) which does not conduct electricity and is therefore used to prevent an electric current from flowing in an undesired direction. Typical insulators are plastics, wood, rubber. A non-electrolyte is specifically a compound.

Insulators in use on a pylon which carries high voltage electric current

Unlike a conductor, a solid electrolyte will not conduct, but if it is melted or made into a solution the current can then pass through the liquid. The electrolysis of hydrochloric acid (Experiment 12.1) decomposed it into its elements; the hydrogen came off at the negative *electrode* (the carbon rod connected to the negative terminal of the battery) which is called the *cathode*. The chlorine was evolved at the positive electrode, called the *anode*, but you did not see so many bubbles because chlorine dissolves in the water of the acid solution. Chlorine gas also came off at the anode when you passed an electric current through the copper chloride solution. The other element in copper chloride, copper, was deposited on the cathode, as a pink coating which you doubtless observed.

If the solid, copper chloride, had been fused by heating, the liquid would have decomposed into copper and chlorine in the same way. However, as you learn more chemistry you will discover that in electrolysis many substances behave differently in solution compared with their behaviour when fused. You found that solid lead iodide would not conduct (Experiment 12.2), but when it was melted current passed and decomposed the compound; the violet vapour of iodine was seen at the anode, and a bead of lead was evolved at the cathode. You may have tested the lead to see if it marked paper.

In Experiment 12.3 you found that a chemical reaction (the magnesium reacting with the acid) caused a flow of electrons which lit the bulb. This was the reverse of electrolysis, and is rather like an exothermic reaction giving out heat. A device which produces electricity like this from a simple chemical reaction is called a primary cell (or voltaic cell). A more convenient primary cell is the dry battery (as used in torches, etc.) in which a different chemical reaction occurs.

How does electricity travel through a fused electrolyte or electrolyte in solution? You will understand how this happens if it is first explained that an electrolyte contains *charged atoms* which are called *ions* (Chapter 9). The number of positive charges and negative charges is always the same, so they balance each other. When the electrolyte is fused or dissolved in water, these ions are free to move about, like the particles in any liquid. If the charged electrodes connected to a battery are now placed in the liquid, the positively charged atoms (positive ions) are attracted to the negatively charged electrode (the cathode) because unlike charges attract each other. Similarly, the negative ions are attracted to the anode (positive electrode). What happens when the ions reach the electrodes? The positive ions collect electrons from the cathode and so become neutral atoms. The negative ions lose their extra electrons at the anode and also become neutral atoms. Now let us take an example to make all this quite clear. Figure 12.6 will help.

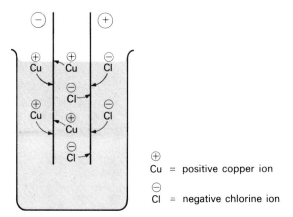

Figure 12.6 Electrolysis of copper chloride

In copper chloride there are positive copper ions and negative chlorine ions. When electrolysis starts, the positive copper ions travel to the cathode, collect electrons, and become ordinary copper atoms. These collect together to form a deposit of copper. The chlorine ions, being negatively charged, go to the anode where they lose their extra electrons and become chlorine atoms. These form into pairs (molecules) and become bubbles of chlorine gas. For every electron taken from the cathode another electron is given to the anode. The result is that the number of electrons leaving the battery for the cathode is exactly equal to the number going back to the battery from the anode, and so there is a continuous flow of current in the circuit. While this happens the compound is decomposed into its elements. Now see if you can describe the electrolysis of fused lead iodide, with a diagram, using the above description as a guide.

The positively charged ions of metals and hydrogen always go to the cathode in electrolysis, and the negative ions of non-metals to the anode. This property of metals enables some of them, for instance chromium, nickel, and silver, to be electroplated onto other metals. The object to be plated is made the cathode during the electrolysis of a solution of a compound containing the metal. For example, in Experiment 12.1(d) a spoon could have been substituted for the carbon cathode and would have become coated with copper. Silver plating and chromium plating are done in the same way. But electroplating is a complicated process and special methods have to be adopted for it to be successful.

Questions

1. Both lead and lead iodide (melted) conduct electricity, but only lead is termed a conductor. Explain why.
2. Classify the following as electrolytes, non-electrolytes, conductors, or insulators:
   (a) sulphuric acid, (b) copper, (c) ethanol, (d) sodium hydroxide, (e) wood, (f) magnesium chloride, (g) sugar, (h) paper, (i) carbon (graphite), (j) lime water.
3. Explain the following terms:
   (a) electrode, (b) electrolysis, (c) cathode, (d) anode, (e) voltameter.
4. Describe what would happen if you passed a direct electric current through fused sodium chloride.
5. Why is it more dangerous to touch wires conducting electricity when your hands are wet?
6. Describe how you could cover a spatula with a coating of copper by using electricity.
7. A boy did the following experiments at home:
   (a) He made a strong solution of sugar and then tried to pass an electric current through it from a car battery, hoping to obtain the hydrogen out of the sugar by fixing an inverted test tube of water over the cathode.
   (b) He tried to decompose hydrochloric acid by dipping two large iron nails into it and connecting the nails by leads to a house power point—an extremely dangerous thing to do.
   (c) Knowing that copper chloride is an electrolyte, he decided to get some copper from it by electricity. He put some of the compound into an evaporating basin and then stuck into it two carbon electrodes connected to a 6 volt battery.
   (d) In order to copper plate a spoon, he connected it to the negative terminal of a 6 volt battery. The positive terminal was connected to a piece of copper wire. Then he dipped the spoon and wire into a solution of copper sulphate.
   (e) He thought he remembered how to make electricity from a chemical reaction, so he connected two wires to a torch bulb in a suitable holder, like the one used at school. To one wire he fixed a piece of zinc and to the other a piece of copper. Then he dipped the zinc and copper into some dilute sulphuric acid.

   You will probably have noticed some of his mistakes. Which of his experiments would have worked? Explain why each of the others would not.
8. Explain how a direct electric current can pass through an electrolyte in solution, using the electrolysis of hydrochloric acid as an example.

13 Treasures from Earth

Discussion and Investigation

Gold, diamonds, and precious stones for jewels have to be dug from the earth. But these are by no means the only treasures hidden in this way. Metals, crude oil, and various other *minerals* such as sulphur are obtained from the ground. Can you think of other substances which are mined?

Some minerals are impure chemical compounds which yield simpler and useful compounds when heated. Examples are chalk, limestone, and marble, which are all forms of the compound calcium carbonate.

Experiment 13.1. To investigate the effect of heating marble (calcium carbonate)

Fix a small marble chip to a piece of iron wire and support the other end of the wire in a retort stand. Place a Bunsen burner under the chip so that the latter is in the hottest part of the flame—just above the light blue cone—for 10 minutes. Allow the chip to cool completely and then examine it. Is it still a hard substance? Put it or pieces of it on a watch glass held in the palm of your hand and add a few drops of water to it from a teat pipette. What happens? What is the explanation for the change in the marble when it was heated? Was a gas evolved? If so, which gas? What do you think is the name of the residue after heating? When water was added to this residue heat was evolved, so a new substance must have been produced by a chemical reaction. Add some more water so that some of the substance dissolves in it. Now add a few drops of an indicator. Is the substance an acid or an alkali? Can you suggest the name of the substance?

Experiment 13.2. To investigate the mineral malachite

Malachite is a greenish coloured rock, but in this experiment it is better to use powdered malachite. From your present knowledge of chemistry you should be able to discover the important compound in the mineral and say what elements are in it. You must decide yourself what tests you are going to do and the apparatus you require. *Some* of the following tests will help you.

(a) The action of heat on the mineral.
(b) Warming the *residue*, after heating, with an alkali solution.
(c) Warming the residue with dilute sulphuric acid.
(d) Testing the approximate solubility of both mineral and residue.
(e) The action of a dilute acid on the mineral.
(f) The electrolysis of any solution obtained in the above tests.
(g) The testing of any gas which you think may be evolved during any of the above tests.

Crude oil or petroleum

You may have already learned about oil from television programmes or books. Where does it come from? How is it obtained? And what is the main process by which the oil is separated into useful substances? Discuss some of these substances and their uses.

Experiment 13.3. Simple fractional distillation of crude oil
Safety precaution
It is essential to heat the crude oil gently, otherwise the evolution of vapour can be violent. Remember that both the oil and the products collected are extremely flammable.

Method
The usual distillation apparatus, with fractionating column and condenser, is not used in this experiment. Instead the simple apparatus of Experiment 3.4 is employed, except that a filter tube (test tube with side arm) replaces the conical flask and the cork contains a thermometer. A little sand is placed in the bottom of the tube to make the oil boil evenly without 'bumping', and then about 3 cm^3 of the oil is added. The bulb of the thermometer should be just opposite the side arm, and the tube should be clamped at an angle so that only its lower end is heated. Heat the oil gently and slowly at first until a few drops of a colourless liquid are collected in the test tube in the beaker. This fraction can be regarded as crude petrol and is removed for later tests. Another test tube is placed in the beaker and the oil heated a little more strongly to collect a slightly off-white fraction, a sort of kerosine (paraffin). In a final test tube a distinctly yellow or brownish fraction is collected by stronger heating for a longer period. This product has a smell like diesel fuel. Test each fraction for viscosity (ease of pouring) and flammability. You should notice a difference in viscosity between the first and last fractions, the former being more 'runny'. For the flammability test, each fraction is placed in an

evaporating basin and lighted with a splint. Observe how easily or otherwise the fraction ignites and how sooty is the flame. If any fraction is difficult to ignite, add a little sand to act as a wick.

Explanation and Further Information

Carbonates

Calcium carbonate, in the forms of limestone (the main rock in most cliffs and sea stones), chalk, and marble, is a very common and useful mineral. Limestone is mixed with clay and heated to make cement. Marble is used for making ornaments and sometimes floors. Calcium carbonate is one of the compounds needed in the manufacture of ordinary glass. When heated, the carbonate decomposes as shown by the following equation:

calcium carbonate + heat → calcium oxide + carbon dioxide

$$CaCO_3 \quad + \text{ heat} \rightarrow \quad CaO \quad + \quad CO_2$$

The white, powdery substance you obtained by heating the marble chip (Experiment 13.1) was calcium oxide, commonly known as quicklime and used as a drying agent since it absorbs water so readily. When water is added to it the calcium oxide becomes calcium hydroxide (slaked lime), and much heat is evolved as you discovered.

$$CaO + H_2O \rightarrow Ca(OH)_2 + \text{heat}$$

Calcium hydroxide is used in mortar, as whitewash, and in making several other chemicals. Its uses as an alkali have been described in Chapter 7. Both calcium oxide and carbon dioxide are produced industrially by heating limestone in large towers called limekilns.

A marble floor at the Taj Mahal in India

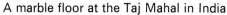

Malachite contains copper carbonate and copper hydroxide, both of which decompose when heated into copper oxide. Carbon dioxide and water are also formed, the latter from the hydroxide. In Experiment 13.2 you should have heated the malachite and obtained moisture and a black residue of copper oxide. From your previous experience with copper carbonate, the colour change would have suggested that the malachite might be impure copper carbonate. The presence of this compound could then have been confirmed by identifying the carbon dioxide evolved and warming some of the black residue with sulphuric acid, when a blue solution of copper sulphate would have been obtained. The equations for these reactions are:

$$CuCO_3 \rightarrow CuO + CO_2$$
$$CuO + H_2SO_4 \rightarrow CuSO_4 + H_2O$$

You may have added an acid to the mineral itself, identified the carbon dioxide evolved, and realized that the resulting blue solution contained a copper salt:

$$CuCO_3 + H_2SO_4 \rightarrow CuSO_4 + H_2O + CO_2$$
$$\text{or} \quad CuCO_3 + 2HCl \rightarrow CuCl_2 + H_2O + CO_2$$

Electrolysis of the blue solution would have shown that it contained copper. From either method you would have concluded that malachite is impure copper carbonate, and that its elements are therefore copper, carbon, and oxygen. In fact, the mineral is what is known as basic copper carbonate ($CuCO_3.Cu(OH)_2$) and so it contains hydrogen as well.

Silicon compounds

Silicon, a non-metal, is the most abundant element after oxygen. *Sand* is impure silicon dioxide and is white when pure. It is used in glass-making and is mixed with cement. The majority of rocks consist of compounds called silicates, several of which, e.g. *asbestos*, *talc*, and *zeolites* for water softening, have important uses. Another form of silicon dioxide is *quartz*, a very hard rock, from which silica glass is made. This has special optical properties and can withstand very high temperatures.

Gypsum

Gypsum is impure calcium sulphate, and when carefully heated is converted into *plaster of Paris* which is used for making stucco, wallboard, and plaster casts for broken limbs, etc. It is often mixed with ordinary plaster and cement. When water is added to plaster of Paris it sets very hard, expanding a little in the process, so the plaster is also used for making casts from moulds.

Coal

Although coal is still used as a heating fuel in some areas, it is much more valuable for the very large number of chemicals that are obtained from it. Some of these are coke, which is a smokeless fuel and a reducing agent for iron and other metal ores, *toluene* (methylbenzene) and other special solvents, *ammonia*, *naphthalene*, and various *gases*.

Crude oil or petroleum

Like coal, petroleum is a complex mixture of many valuable substances. In the oil refinery, the petroleum is first separated into

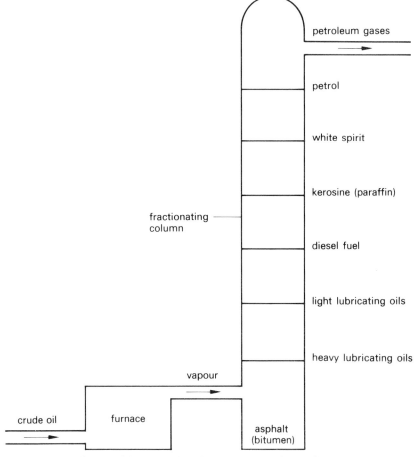

Figure 13.1 Fractional distillation of petroleum

certain liquids and gases by fractional distillation. The residue from this process is *asphalt or bitumen* which is used for road-making, roof felt, etc. The liquids, which are mixtures of many compounds called hydrocarbons, are crude forms of *petrol, white spirit, kerosine or paraffin, jet fuel, diesel engine fuel, fuel oil,* and *lubricating oils.* These products all have to be carefully processed, and in this way several by-products such as *paraffin wax* (candle wax) are obtained. Many important substances, such as *synthetic rubber, alcohols,* and *plastics* are made from the petroleum gases, and they are also used as fuels (*bottled gas*).

Before the oil enters the tall fractionating tower it is vapourized by being heated in a special furnace. The vapours pass up the tower, which becomes cooler nearer the top, and condense to liquids on trays inside the tower. Liquids like crude petrol, which have low boiling points, collect on the higher, cooler trays, and those with high boiling points, such as lubricating oils, on the hotter trays nearer the bottom of the tower. The liquids are transferred from the various trays to special processing plants in which the finished products are made. Figure 13.1 shows a petroleum distillation plant diagramatically.

Questions

1. What are the chemical names of the following:
   (a) limestone, (b) marble, (c) gypsum, (d) quicklime, (e) slaked lime?
2. Mention one important use of each substance listed in Question 1.
3. What happens when water is added to the compound obtained by heating limestone? Name the compounds involved in the reaction and write the equation.
4. From what minerals can the following be made:
   (a) cement, (b) quicklime, (c) carbon dioxide, (d) plaster of Paris, (e) glass?
5. Write about half a page on the importance of silicon compounds.
6. Describe, with equations, how you investigated the mineral malachite.
7. Explain why coal is more important as a useful mineral than as a fuel.
8. Describe briefly the industrial separation of crude oil (petroleum) into its main products.
9. Mention briefly the mineral source and one main use of the following:

(a) asphalt (bitumen), (b) ammonia, (c) coke, (d) bottled gas, (e) paraffin wax.

10. A white compound reacted with hydrochloric acid, forming three products of which one was a salt used in soldering. When heated, the compound evolved a gas which turned lime water milky; the residue was yellow at first but turned white when cool. Name the compound and its elements.

14 Bright and Buried

Discussion and Investigation

We are all familiar with metals because most of the well known ones are strong, shiny materials which are so essential in the construction of such a variety of things, from huge ships, planes, and bridges to small objects like radios, watches, and rings. Metals, in common with many other useful substances, come from the ground. Few are bright when they are mined because they are combined with non-metals (such as oxygen and sulphur) in compounds called ores. An obvious exception is gold which, like platinum and a few other metals, is found in the free state (i.e. as an element). But the other metals are bright and shiny too when they have been separated in chemical works from their ores.

You will discover more about the chemical properties of metals in this chapter, and how metals are obtained from their compounds. You have probably already learned in physics about several of the physical properties of metals. A metal is a good conductor of electricity, for example. How do metals and non-metals differ in strength, density, and lustre (shininess)? Why do cooking pans, electric irons, etc., have plastic or wooden handles? Do you think that a bell could be made from a non-metal element? Can a non-metal be used for making wire?

Sometimes, metals are mixed together. The dentist does this with two metals to make fillings for your teeth. Do you know the name for a mixture of metals? Are the metals brass, bronze, and stainless steel elements or mixtures of elements?

Discuss the uses of some well known metals such as iron, copper, aluminium, lead. You may know that steel is a toughened form of iron made by mixing a little carbon or other substances with the iron. What are some of the advantages and disadvantages of these metals?

When studying hydrogen you found that it is a reducing agent and can reduce copper oxide to copper by removing its oxygen:

$$\text{copper oxide} + \text{hydrogen} \rightarrow \text{copper} + \text{water}$$
$$CuO + H_2 \rightarrow Cu + H_2O$$

Many metal ores are oxides, and hydrogen can reduce some of them. A better way to obtain many metals from their ores is to use carbon, in the form of charcoal or coke. Carbon is a stronger and cheaper reducing agent than hydrogen.

The following experiments should be done in a fume cupboard.

Experiment 14.1. To obtain lead from lead oxide (litharge)

Mix two spatula measures of the yellow lead oxide with the same quantity of powdered charcoal and place the mixture in a crucible lid supported on a wire gauze and tripod stand. Heat the mixture strongly for about five minutes, holding the Bunsen burner in your hand so that the tip of the inner blue flame touches the mixture. Examine the mixture carefully when it is cool. Are there any lead particles? A more effective method is described in Demonstration 14.2 which your teacher will probably show you.

Demonstration 14.2. Extraction of metals from their oxides

Make a small hole in a charcoal block and press a spatula measure of lead oxide (litharge) into it, together with some of the charcoal powder. Heat the oxide strongly with the Bunsen flame, using a blow pipe to increase the temperature, until a fairly large drop of molten lead is obtained. Demonstrate the marking of paper with the piece of lead when it is cool. Repeat the experiment with red iron oxide and zinc oxide.

In Experiment 7.7 you found that most metals react with dilute acids and evolve hydrogen. You will have noted that the metals you tested varied in the speed with which they reacted with the acid, magnesium being the best or most reactive metal. Perhaps active metals, like magnesium, react with water too? If certain metals do react with water we should probably expect a less vigorous reaction than with acids. In the following experiment you are going to find out which of several metals react with water, one of them, calcium, being too reactive for you to use in Experiment 7.7.

Experiment 14.3. To find out which metals react with water

(a) *Calcium*

Place a few calcium turnings in half a test tube of water. Bubbles of a gas are evolved. Test the gas as you did in Experiment 7.7. Can you identify it? The water turns a milky colour, so a white solid has also been formed. Some of this solid dissolves in the water. Add a few drops of phenolphthalein or Universal Indicator. What type of compound is the white substance? What is its name?

(b) *Magnesium*

Polish a short length of magnesium ribbon with emery paper to remove the tarnish, and place it in a small beaker of cold water. Are there any signs of a chemical reaction? Now put the magnesium in a beaker of very hot but *not boiling* water. What do you see? What happens to the magnesium? Why does it float to the surface? The reaction is slow, but if an inverted test tube of hot water is placed over the magnesium enough gas can eventually be collected to test as above. Add a few drops of phenolphthalein to the water. In due course a pale pink colour indicates the presence of a weak alkali. What is its name?

Other metals

Repeat Experiment 14.3 (b) to test aluminium, zinc, and iron, using emery paper if necessary to polish the metal.

Sodium is a very reactive metal and for this reason it has to be kept under oil so that no air can reach it. What would happen to the metal if it were left exposed to the air? How do you think it will react with water, compared with calcium and the other metals?

Demonstration 14.4. **To show the reaction of sodium with water**

A very small piece of the metal is held in tongs, wiped with filter paper, and dropped into a large beaker or bowl of water. A sheet of perspex is held over the water in case the sodium reacts too violently.

Questions for you to ask yourself during the experiment:

1. Why does the piece of sodium become a shiny silver globule when it is put on the surface of the water?
2. What enables it to dart about on the surface, like a miniature hovercraft?
3. If hydrogen is evolved, how could this be demonstrated?
4. Can you name the compound which is formed by the reaction, and say how you would show its presence?

Perhaps the metals which react with water would also react with steam. Would you expect the reaction to be more vigorous, less, or the same?

Demonstration 14.5. To show the reaction of magnesium with steam

(It is advisable to use a safety screen and protective equipment and to ensure that the flask has no cracks.) Fit a round-bottomed flask with a cork and right-angled glass tube connected to a fairly long piece of rubber tubing as delivery tube. The cork also holds a combustion spoon to which a small coil of magnesium ribbon is attached. A large test tube is filled with water, inverted in a trough or small bowl of water, and clamped in a retort stand. Fill the flask with steam by boiling a small quantity of water in it. Then, with the end of the rubber delivery tube firmly in the test tube, light the magnesium and quickly plunge it into the flask so that the cork is in the neck. Here are some questions for you concerning the demonstration.

1. How can the magnesium continue to burn in the steam, since there is no air in the flask?
2. What is the gas collected in the test tube?
3. What is the white compound formed?
4. Is it soluble in water? If it reacts with water, what compound would be formed, and how would you identify it?
5. Did the magnesium react more vigorously with steam than with water? Explain your answer.

Experiment 14.6. To find out whether other metals react with steam

Put a little wet sand in the bottom of a test tube and then clamp the tube horizontally in a retort stand. Place about two spatula measures of fine iron filings half way down the test tube. Fit a cork and delivery tube to the test tube so that if hydrogen is

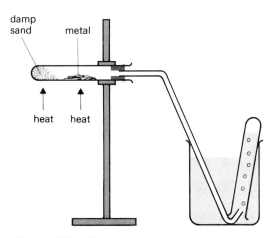

Figure 14.1 Reaction of metals with steam

evolved it can be collected in a test tube over water (see Figure 14.1). Heat the iron filings nearly to redness in a hot Bunsen flame, and then heat the sand gently in order to produce the steam. It is better to do this with a second Bunsen burner so that the iron can be heated without interruption. It is an important precaution to *disconnect the delivery tube as soon as you stop heating*. Why is this? Is hydrogen evolved? What happens to the iron? Can an alkali be formed from any product of the reaction? Now repeat the experiment with lead, zinc, copper, and aluminium, all in the form of powders, being careful to observe the above *safety precautions*.

If you consider carefully the results of the experiments you have done or seen demonstrated, you will be able to make a list of the metals which reacted with water or steam, showing the approximate order of their chemical reactivity. Does the order agree, as regards zinc, iron, and magnesium, with the results of your experiments with metals and acids (Experiment 7.7)?

Explanation and Further Information

Most elements are metals, but many of them are not well known and their compounds occur in only small quantities. The more abundant or better known metals are listed at the end of Chapter 5.

An alloy is a compound or a mixture of two or more metals which has special properties. A few alloys also contain small quantities of non-metals. Examples of alloys are brass (copper and zinc), bronze (copper and tin), solder (lead and tin), and stainless steel (iron, chromium, and carbon). Magnalium is an alloy used in the construction of aeroplanes, boats, and cars. It contains aluminium, magnesium, and copper, and is lighter, harder, and stronger than aluminium.

Physical properties of metals

Metals are usually much stronger than solid non-metals, which are often brittle or soft. Metals are good conductors of heat and electricity, and most have higher densities than non-metals. Carbon (in the form of graphite) is the only non-metal which conducts electricity.

Unlike non-metals, metals

    (a) are ductile, i.e. can be drawn out into wire;

    (b) are malleable, i.e. can be hammered into thin sheets;

(c) have a metallic ring when they are hit;
(d) have a metallic lustre.

Chemical properties of metals

The more important properties, several of which you have already learned, are listed below.

Metals
(a) usually react with most dilute acids to evolve hydrogen;
(b) combine with oxygen to form oxides which react with acids to form salts and water. Some of these oxides are alkaline oxides (Chapter 7);
(c) combine with other non-metals, such as sulphur and chlorine;
(d) form salts by replacing hydrogen in acids;
(e) decompose water or steam if sufficiently reactive;
(f) are deposited on the cathode (negative electrode) when their compounds are decomposed by electrolysis (Chapter 12);
(g) form positive ions by their atoms losing electrons.

Reaction of metals with water

Metals which can decompose water or steam are, in decreasing order of reactivity, potassium, sodium, calcium, magnesium, zinc, and iron. With water they produce hydrogen and alkalis (soluble metal hydroxides like calcium hydroxide) and with steam, hydrogen and the metal oxides (e.g. magnesium oxide) are formed. Aluminium does not react with steam (see page 130).

Extraction of metals from their ores

Most metal ores are impure oxides or sulphides. The ores are separated from earth, etc., and if they are sulphides they are

Letting molten iron flow out of a blast furnace at Newport, after it has been separated from the ore

converted into oxides. The oxides are reduced by heating with coke (impure carbon). Iron, tin, lead, and zinc are obtained in this way. The main chemical reaction is the same as the one you saw in Demonstration 14.2, e.g.

$$PbO + C \rightarrow Pb + CO \text{ (carbon monoxide)}$$

The carbon monoxide becomes oxidized to carbon dioxide. It is difficult to obtain zinc by this method in the laboratory because a very high temperature is needed.

You will not be surprised to learn that ordinary reducing agents, such as carbon, are not strong enough to obtain the very reactive metals from their compounds. This can be done by using special reducing agents (see Chapter 15), but in fact the quicker and frequently cheaper method of electrolysis is used, and the metals sodium, calcium, magnesium, and aluminium are extracted in this way. The metals are deposited on the cathode, as in Experiment 12.2.

Questions

1. Compare the physical properties of metals and non-metals.
2. State which of the following metals react with steam:
   (a) zinc, (b) lead, (c) copper, (d) magnesium, and (e) iron. What are the products when a metal reacts with steam?
3. Describe the reaction of sodium with water and explain how you would identify the products.
4. Explain the terms (a) ore, (b) reducing agent, (c) ductile, (d) malleable, and (e) alloy.
5. Describe, with a diagram, an experiment to show the reaction of magnesium with steam.
6. Name two alloys and their uses.
7. Describe briefly an experiment by which a sizeable piece of lead can be obtained from lead oxide.
8. Describe four chemical properties of metals by giving examples of the chemical reactions concerned.
9. Which of the following metals can be obtained from their oxides by using coke as a reducing agent?: (a) lead, (b) aluminium, (c) magnesium, (d) iron, and (e) copper. What method could be adopted for those metals which cannot be extracted in this way?
10. Name
    (a) a metal which is found in the free state;
    (b) a light alloy used in aeroplanes;
    (c) a non-metal which conducts electricity;
    (d) a metal which has to be kept under oil to prevent it from reacting with substances in the air;
    (e) a metal which is extracted from its compound by electrolysis because ordinary reducing agents are unsuitable.

15 Form Order

Discussion and Investigation

Figure 15.1

Metals

Your experiments involving the reaction of metals with both acids and water (or steam) have enabled you to list them in what you considered to be their order of chemical reactivity. You may not have put every metal in its correct order because two metals may behave similarly in a particular reaction, or in certain reaction conditions. Also the strength or vigour of a particular chemical reaction is not necessarily an indication of the reactive power of an element. But an approximate 'batting order' can be made from your data, and the positions of First, Second, Third, and Last easily decided:

1st	*2nd*	*3rd* . . . . *Last*
Sodium	Calcium	Magnesium . . . . Copper

When it is not easy to decide between one metal and another, the results of other experiments can be considered. For example, metals can be compared by considering their eagerness to combine with oxygen, their 'affinity for oxygen' as chemists call this property.

126

In your study of air and oxygen you would have noted that the metals sodium, calcium, magnesium, and zinc (powder) combine quickly with oxygen, and that others, such as lead and copper, do so much more slowly. Since zinc has a greater affinity for oxygen than copper has, do you think that it could reduce copper oxide to copper in the way that carbon or hydrogen can? If zinc can act as a reducing agent like this, the reaction could be represented by the following equation:

$$Zn + CuO \rightarrow ZnO + Cu$$

What would be the colour changes? If this reaction is possible there must be other metals which could also reduce copper oxide. Do you think that copper could reduce zinc oxide? From your present knowledge of the differences in reactivity of metals you should be able to deduce an apparent rule or law for predicting whether a certain metal will reduce a particular metal oxide. In the next experiment you can find out whether your hypothesis was correct.

Experiment 15.1. Reduction of metal oxides by metals

Thoroughly mix together one spatula measure of the metal powder and one of the oxide, as indicated in the list below, and place the mixture on a crucible lid supported on a wire gauze and tripod stand. Heat the mixture directly with a Bunsen flame, holding the burner in your hand. You should wear goggles and gloves and keep your face away from the apparatus, since some of the reactions are quite vigorous and a few hot particles may spurt a little. Note carefully any sign of a chemical reaction such as a sudden red glow spreading through the powder or permanent colour change. By recording whether or not reactions take place you will be able to analyse your results and produce a reactivity order for the metals concerned.

The pairs of chemicals you should use are:
(a) copper oxide and zinc powder;
(b) magnesium oxide and zinc powder;
(c) zinc oxide and iron filings (fine);
(d) red iron oxide and zinc powder;
(e) lead oxide (litharge) and iron filings (fine);
(f) lead oxide (litharge) and copper powder.

Write word equations for the reactions.

Demonstration 15.2

Repeat Experiment 15.1, using magnesium powder and copper oxide, but place the Bunsen burner on the bench under the mixture which should be in a crucible on a pipe-clay triangle. Hold a perspex safety screen over the reactants to catch the spurting products.

A well known practical application of the metal/metal oxide type of reaction is the Thermit process. This consists of the reduction of iron oxide by aluminium powder to produce molten iron. The temperature of the reaction is 3000 °C. Thermit incendiary bombs were used in the second world war; water had very little effect on them. Aluminium powder is used similarly for the production of small quantities of other metals such as chromium.

Demonstration 15.3. The Thermit reaction
(This demonstration should be done *out of doors*.)
Mix together about 5 g of red iron oxide and an equal volume of fine aluminium powder. Both should be thoroughly dry. Put the mixture in a silica or fire clay crucible and place this in a bucket or tin of sand. Now make a small cavity in the surface of the mixture, and put into it about two spatula measures of a mixture of barium peroxide and magnesium powder. Stick into this mixture a length of magnesium ribbon to act as a fuse. Light the magnesium and then quickly stand well away from the apparatus until the violent reaction is over. When the crucible is cool, tip out the contents on to a sand tray.
Questions to consider concerning the demonstration:
1. What is the purpose of the mixture of magnesium powder and barium peroxide?
2. What is the substance that glows red in the crucible after the reaction?
3. How would you identify the piece of iron produced?

You have learned about replacement reactions (Chapter 10) and will now be able to explain them. All the metals (except copper) which you tested with dilute acids were able to displace hydrogen from the acids. This must mean that they are more reactive than hydrogen. Does it also indicate that copper is less reactive? Do you know of a reaction in which hydrogen can 'beat' copper in a competition for oxygen?

Non-metals
Evidently hydrogen can be placed in a reactivity order of the metals. See if you can find the right place for carbon, inserting it

between two of the metals to show that it is chemically stronger than some and weaker than others. It will help if you read through Chapter 14. The following experiment will enable you to decide approximately how high to place it.

Experiment 15.4

Make a gas jar of carbon dioxide as in Experiment 8.5. Hold one end of a short length of magnesium ribbon with tongs and light the other end in a Bunsen flame. As soon as the magnesium begins to burn, quickly remove the gas jar cover and plunge the magnesium into the gas. Does the metal continue to burn? When the reaction is over, examine the inside of the jar. What is the white powder? And the bits of black substance on the sides of the jar? Write the word equation for the reaction. Which is more reactive, magnesium or carbon?

The non-metals can themselves be compared concerning reactivity. You already know that carbon is a stronger reducing agent than hydrogen. Here is an experiment to show which is the more reactive, chlorine (one of the elements in hydrochloric acid and in sodium chloride (salt)) or iodine (a crystalline non-metal which is brown when precipitated).

Demonstration 15.5

Add a little dilute hydrochloric acid to some sodium hypochlorite in a test tube, to produce a small quantity of chlorine gas. Hold over the mouth of the test tube a filter paper moistened with potassium iodide solution. What happens? Which of the two elements is the more reactive? Explain your answer.

Explanation and Further Information

Metals

In order to find the reactivity order of the metals you had to carry out several experiments because a single experiment does not always give a clear indication of reactivity difference between two metals. Before we consider the results of your experiments, it will be helpful to outline some of the difficulties encountered in the three types of reaction.

Metals and dilute acids

Although the Thermit reaction shows clearly that aluminium is more reactive than iron, just the opposite appears to be the case when the metals are placed in dilute acids. This is because aluminium has a hard, thin oxide film on its surface, and this slows down the reaction with acids.

Metals and steam

The oxide layer on aluminium prevents its reaction with steam. With zinc and iron it is not easy to decide which metal is the more reactive, and lead does not react with steam at ordinary temperatures.

Metals and metal oxides

This is a more reliable method for determining reactivity order, for even a slow reaction indicates that the metal reducing agent is more reactive than the metal in the oxide. If there is no reaction, the reducing agent is less reactive because it cannot overcome the power of the metal in the oxide to hold on to its oxygen. You may ask why the absence of a reaction does not indicate that both metals are equally reactive, like two equally powerful tug-of-war teams. But even in tug-of-war competitions, one team eventually has to yield, and it is the same with chemistry. There are no elements of exactly the same chemical power or reactivity, and you will in due course do experiments which show this. But even the metal/metal oxide reaction has its limitations. It is sometimes difficult to be sure that a reaction has occurred, and some metals, e.g. aluminium, require more activation energy (as heat) than others to make them react. In the Thermit reaction the aluminium has to be strongly heated by the exothermic reaction between the barium peroxide and magnesium.

Reactivity order

The complete order of chemical reactivity—called the 'Electrochemical Series' or 'Activity Series'—for all the common metals is shown below:

1	2	3	4	5	6	7	8	9	10	11	12	13	14
K	Ca	Na	Mg	Al	Zn	Fe	Sn	Pb	H	Cu	Hg	Ag	Au

← *More reactive*   *Less reactive* →

From the experiments and demonstrations, you will have realized that the vigour of the reaction between a metal and a metal oxide depends on how close the two metals are in reactivity. For example, there is a good reaction between zinc and copper oxide, and an even better one between magnesium and copper oxide. But the reaction between lead and copper oxide is slow because these two metals are near each other in chemical reactivity.

You will notice that hydrogen has been included with the metals in the series. There are good reasons for this which you will understand better when you have learned more chemistry. In chemical reactions hydrogen often behaves like a metal; in combining with non-metals, for example. A glance at some formulae of hydrogen compounds and metal compounds makes this clear:

HCl H_2SO_4 H_2O
$NaCl$ Ag_2SO_4 K_2O

All metals above hydrogen in the Series displace it from dilute acids, but metals which are less reactive than hydrogen, like copper, cannot do this. A more reactive metal can displace a less reactive one from a solution of its salt (Chapter 10, replacement reactions) and can reduce the oxide of a less reactive metal, as you have discovered.

It may surprise you that calcium is placed before sodium. You will understand why this is so at a later stage in your chemistry course.

You were asked to consider the position of carbon in the Series. Experiment 15.4 showed that it was less reactive than magnesium, and in previous experiments it was found to be more reactive than iron, lead, and copper. At high temperatures it is also more reactive than zinc. But you would be unable to reduce aluminium oxide with carbon, so its correct place among the metals is between aluminium and zinc.

Now you must check your recorded results of the experiments and demonstrations.

Experiment 15.6
(a) Zinc is more reactive than copper, so it reduced the black copper oxide to pink copper and was itself oxidized to white zinc oxide which is yellow when hot. The reaction may not have been completed, in which case some of the reactants, e.g. copper oxide, would remain.
(b) No reaction. Magnesium is more reactive than zinc.
(c) No reaction. Zinc is more reactive than iron.
(d) A slow reaction because zinc is not very much more reactive than iron. White zinc oxide and silver grey iron are formed.
(e) A slow reaction because iron is a little more reactive than lead. Iron oxide is formed and the yellow lead oxide is reduced to lead.

(f) No reaction. Copper is less reactive than lead.
From the above results, the order of reactivity is:

1	2	3	4	5
Mg	Zn	Fe	Pb	Cu

Demonstration 15.7

A very vigorous reaction occurs because magnesium is so much more reactive than copper. It is oxidized to white magnesium oxide and the black copper oxide is reduced to pink copper.

Demonstration 15.8

The burning piece of magnesium evolves enough heat to cause the barium peroxide and magnesium powder to react. The magnesium reduces the barium peroxide to barium oxide (BaO) and is oxidized to magnesium oxide. The heat given out from this reaction is enough to set off the main reaction in which aluminium is oxidized to white aluminium oxide and the red iron oxide reduced to iron. The high temperature (3000 °C) causes the iron to be molten when first formed. It can be identified by its magnetic properties, but a better way is to warm a piece of it with hydrochloric acid. If an alkali is added to the resulting solution of iron chloride a green precipitate of iron hydroxide is formed.

Experiment 15.9

The magnesium burns in the carbon dioxide by removing the oxygen from (reducing) it. The products are therefore white magnesium oxide and black carbon.

Non-metals

Non-metals can be inserted between the metals in the Activity Series, depending on their reactivity. Examples are carbon and hydrogen (page 131). It is also interesting to compare the reactivity of non-metals together, especially when they have similar properties. The non-metal elements fluorine, chlorine, bromine, iodine, and astatine are closely related in their properties, forming a sort of family of elements (page 133). In Experiment 15.5 you found that chlorine is more reactive than iodine because it displaced iodine (the brown stain on the filter paper) from the salt potassium iodide. The order of reactivity of these elements is as they are listed above.

The Periodic Classification of the elements

The relative reactivity of the elements, and many other chemical properties, are shown in the Periodic Classification Table. In this table the elements are arranged in the order of their *atomic numbers* (the number of electrons each atom of an element has), and this results in vertical columns (called Groups) of elements having similar properties. Two of these Groups are very well known. One is the *halogens* (Group 7), the family of elements mentioned on page 132. The word 'halogen' means 'salt maker', for these elements all combine directly with metals to form salts such as sodium chloride, silver bromide, and lead iodide. The other is the group known as the *alkali metals* (Group 1). These are lithium, sodium, potassium, rubidium, caesium, and francium, and they all react with water to form alkalis (e.g. sodium hydroxide). The table is the most important classification in chemistry and is complicated. The beginning of it, the first eighteen elements, is shown below (Figure 15.2) to give you some idea of its form. You will notice that by arranging the elements in this way there is an interval or period of eight elements between those having similar properties (e.g., lithium and sodium, fluorine and chlorine). This interesting relationship was first discovered in 1864 by the English scientist, Newlands, who compared this chemical period to an octave in music. Note that hydrogen, which behaves chemically like both a metal and a non-metal, is an exception to this eighth-element characteristic.

Period	Group 1	Group 2	Group 3	Group 4	Group 5	Group 6	Group 7	Group 0
1	H (1)							He (2)
2	Li (3)	Be (4)	B (5)	C (6)	N (7)	O (8)	F (9)	Ne (10)
3	Na (11)	Mg (12)	Al (13)	Si (14)	·P (15)	S (16)	Cl (17)	Ar (18)

Figure 15.2 Periodic Table (first eighteen elements)

Note   Symbols of the common elements will be found on page 38. Symbols of other elements in the above table are:

Lithium	Li		Helium	He
Beryllium	Be		Neon	Ne
Boron	B		Argon	Ar
Fluorine	F			

Questions

1. Explain briefly why it is necessary to consider the results of several chemical reactions in order to investigate the reactivity of metals.
2. What is meant by the statement 'This metal has a great affinity for oxygen'? Give an example of such a metal and describe briefly how you would show that it possessed this property in contrast with another metal.
3. From the following experimental results determine the order of reactivity of the metals A, B, C, and D, putting the most reactive metal first. Explain your reasoning.
   (a) C's oxide is not reduced by A.
   (b) When B is heated with A's oxide a red glow spreads through the mixture.
   (c) C reduces D's oxide in a very vigorous manner.
   (d) When C's oxide is heated with B nothing happens.
   (e) When B's oxide is heated with D nothing happens.
   (f) C reduces A's oxide fairly vigorously.
4. Describe what would happen if the following oxides and metals (as powders) were heated together:
   (a) copper oxide and lead; (b) aluminium oxide and zinc; (c) magnesium oxide and copper; (d) copper oxide and magnesium; (e) lead oxide and iron.
   Would it make any difference in the last reaction (e) if the iron was in the form of iron nails? Explain your answer.
5. Why is hydrogen included in the Electrochemical Series of the metals? Where would you place carbon in the Series?
6. Arrange the following in the order of their chemical reactivity, putting the most reactive first:
   copper, sodium, mercury, lead, silver.
7. John's friend said that he could easily obtain metals from chemicals and that his method was better than extracting metals from their ores. To prove his point he added that he would show John how to obtain copper, lead, and iron, and this is what he did:
   (a) For copper, he put some blue copper sulphate in a beaker and added some iron nails.
   (b) To obtain lead, he filled a beaker with lead nitrate solution and added some pieces of zinc.
   (c) For the iron, he put pieces of lead in iron chloride solution.
   Was his method better than the usual one? Explain your answer. Comment on his efforts to obtain each metal.
8. What happens if:
   (a) A piece of copper foil is placed in silver nitrate solution?
   (b) Chlorine is bubbled into a solution of potassium iodide?

   (c) Bromine is added to sodium chloride solution?

   (d) Carbon and mercury oxide are heated together?

   (e) Iron filings are mixed with aluminium oxide and then strongly heated by the reaction between barium peroxide and magnesium powder?

9. Describe how the metal chromium can be obtained by the Thermit reaction.

10. Describe the basis of the Periodic Table and give examples of how it shows elements possessing similar properties.

16 Chemical Equations

Discussion and Investigation

Do you find mathematical equations difficult? Most pupils like the simple kind, and some even enjoy the intricacies of quadratic equations. You could solve this equation in your head:

$$2x + 2 = 6$$

And you can doubtless write it in words:

Twice something plus two is equal to six.

Chemical equations are no harder than simple equations. Is this one hard?:

$$A + B \rightarrow AB$$

In words, substance A reacts chemically with substance B and they produce a new substance AB. As you probably know the chemical symbols of the common elements and the formulae of several well known compounds, we can write a proper chemical equation:

$$Fe + S \rightarrow FeS \tag{1}$$

You may know this reaction. The equation tells us that the element iron combines chemically with the element sulphur to form the compound iron sulphide.

Symbols and formulae

When formulae are more complicated (e.g. H_2O, CO_2), the equation is a little harder to write, so before we go any further it is as well to recall what you learnt about symbols and formulae in Chapter 5 and expand this a little. A *symbol* represents one atom of the *element* concerned. If it is necessary to write two or more atoms of an element, a large figure is placed in front of the symbol, e.g. 2 Zn (two atoms of zinc), 2 H (two separate atoms of hydrogen). When a symbol is followed by a small number, the meaning is one molecule of that number of atoms of the element. For example, H_2 signifies a molecule of hydrogen (consisting of two atoms). A *formula* indicates the ratio of the different atoms in a *compound*, so in iron sulphide (FeS) there are equal numbers of iron and sulphur atoms, but water (H_2O) contains twice as many hydrogen atoms as oxygen atoms. In aluminium oxide (Al_2O_3) there are two aluminium atoms for every three oxygen atoms. It is because atoms always combine together in specific numbers like this when they form compounds that 'compounds consist of elements combined together in definite weight proportions'.

Some simple equations

We can now consider some more equations, but first remember that the particles in many gases are molecules consisting of two atoms, so the symbols for such gases are written as O_2, H_2, N_2, etc., meaning one molecule of the gas (Chapter 9). More than one molecule is shown by placing a large number in front of the symbol; for example, '2 O_2' means 'two molecules of oxygen'. This also applies to compounds, so '2 CO_2' indicates two molecules of carbon dioxide (each molecule consisting of one atom of carbon and two of oxygen). When you burnt carbon in oxygen, carbon dioxide was produced:

$$C + O_2 \rightarrow CO_2 \qquad (2)$$

This equation takes much longer to write in words:

One atom of carbon combines with one molecule of oxygen to form one molecule of carbon dioxide.

So equations can be regarded as a sort of chemists' shorthand. But you will learn that they are a great deal more than this. Now see if you can write down the following equation in words:

$$H_2 + S \rightarrow H_2S \qquad (3)$$

When you have done that, write proper equations from the following word equations:

One atom of sulphur combines with one molecule of oxygen, producing one molecule of sulphur dioxide.

One molecule of carbon dioxide reacts with one atom of carbon to form two molecules of carbon monoxide (CO).

Equations as recipes

A recipe for concrete is to mix together one bucket of cement, two buckets of sand, two of small stones or chips, and, say, half a bucket of water. These ingredients react together to form a hard mass of concrete. If we invent symbols for them we can write the recipe as an equation. Let C = cement, S = sand, Ch = chips, W = water, and Cn = concrete. Then the equation is:

$$C + 2S + 2Ch + \tfrac{1}{2}W \rightarrow 5\tfrac{1}{2}Cn \qquad (4)$$

What does each symbol indicate in addition to the name of the ingredient?

You do not have to be much of a mathematician to remove the fractions in this equation and obtain the following one:

$$2C + 4S + 4Ch + W \rightarrow 11Cn \qquad (5)$$

Has the recipe for concrete altered?

If we wanted to make a lot of concrete we could use this equation for the same recipe:

$$200C + 400S + 400Ch + 100W \rightarrow 1100Cn \qquad (6)$$

A chemical equation represents the smallest quantities (e.g. atoms and molecules) of the substances in the reaction, like equation (5) above. Note that fractions are not usually used in chemical equations. (Why?) You will appreciate that in a typical laboratory reaction the amounts of substances reacting together (several grams perhaps) consist of millions of atoms, etc.

Equations and the law of conservation of matter

When concrete is made are any of the ingredients lost? Are the number of buckets of ingredients equal to the number of buckets of concrete made? In Chapter 10 you learned that nothing is gained or lost in a chemical reaction, and one which was considered was the decomposition of hydrated copper sulphate into anhydrous copper sulphate and water. The equation is:

$$CuSO_4.5H_2O + heat \rightleftharpoons CuSO_4 + 5H_2O$$

You will notice that nothing is lost or gained whether the reaction proceeds from left to right or the other way, that there are exactly the same number of copper, sulphur, hydrogen, and oxygen atoms on each side of the arrows. You may wonder about the formula for hydrated copper sulphate. All hydrates contain water of crystallization which is loosely combined and fairly easily removed by heat; their formulae are always written like this one, a full stop being placed between the main compound and the water molecules. Other examples are gypsum (calcium sulphate) ($CaSO_4.2H_2O$) and washing soda (sodium carbonate) ($Na_2CO_3.10H_2O$).

Look back at the equations (2) and (3) on page 137. Count up the atoms and see if any have been gained or lost. Now you can complete the following sentence:

> In a chemical reaction the number of atoms of each kind in the reactants is always equal to.....................

Balancing equations

This exact balance between reactants and products is an essential requirement in equations. If the atoms do not add up correctly on both sides of the arrow the equation must be wrong. For example, this equation is wrong:

$$H_2 + O_2 \rightarrow H_2O \qquad (7)$$

The symbols and formulae are correct, but the oxygen atoms do not balance. How do we balance them? If the formula of water were altered to H_2O_2 the equation would be balanced, but would it be correct? Is water H_2O_2? Perhaps you know what the compound H_2O_2 is. The only way to balance the equation is to have *two* molecules of hydrogen and *two* molecules of water:

$$2H_2 + O_2 \rightarrow 2H_2O \qquad (8)$$

And this is what in fact happens when two *volumes* of hydrogen combine with one *volume* of oxygen. Two *volumes* of steam (and perhaps a big bang) are produced. On the smallest scale, two *molecules* of hydrogen react with one of oxygen, forming two *molecules* of steam, as the equation shows.

Here are some easy equations for you to balance. They are skeleton equations, i.e. the symbols and formulae are correct, but the equations are unbalanced. Copy them into your rough book and then balance them. Remember that formulae cannot be changed (why?):

$$Zn + HCl \rightarrow ZnCl_2 + H_2 \qquad (9)$$
$$NaOH + H_2SO_4 \rightarrow Na_2SO_4 + H_2O \qquad (10)$$
$$FeS + HCl \rightarrow FeCl_2 + H_2S \qquad (11)$$
$$H_2S + O_2 \rightarrow H_2O + S \qquad (12)$$

Equations with more complicated formulae

You have met with formulae like $Ca(OH)_2$, calcium hydroxide. These sometimes make equations a little harder to balance, but the method is just the same. A group of atoms is enclosed in brackets, with a small figure outside, to show that a certain number of groups is combined with an element. In this case, two hydroxyl (OH) groups are combined with one calcium atom. Such groups are called *radicals* and often do not change in chemical reactions. You will remember a replacement reaction in which the sulphate radical (SO_4) did not change:

$$Zn + CuSO_4 \rightarrow ZnSO_4 + Cu$$

Other examples of formulae of this type are magnesium nitrate ($Mg(NO_3)_2$ and aluminium sulphate ($Al_2(SO_4)_3$). Now try balancing these equations:

$$Mg(OH)_2 + HNO_3 \rightarrow Mg(NO_3)_2 + H_2O \qquad (13)$$
$$Ca(OH)_2 + HCl \rightarrow CaCl_2 + H_2O \qquad (14)$$
$$Na + H_2O \rightarrow NaOH + H_2 \qquad (15)$$
$$Al(OH)_3 + H_2SO_4 \rightarrow Al_2(SO_4)_3 + H_2O \qquad (16)$$

After you have balanced these skeleton equations, write the word equation (equations (9) to (16)) and then revise any experiments in which the reactions occurred.

Remember that you cannot write an equation unless you know the formulae of the compounds concerned. Until you learn more chemistry you will have to look up the formulae if you do not know them, and a list of common ones is on page 146. It helps if you write the word equation first and then put the correct symbols and formulae under the names to obtain the skeleton equation. The final step is to balance this equation, as we have seen.

Never think of equations as just exercises in theoretical chemistry. They represent what actually happens in an experiment involving a chemical reaction. You will understand this better when you have read the next part of the chapter.

Summary and Further Information

First check your balancing of equations (9) to (16). The balanced equations are:

$$Zn + 2HCl \rightarrow ZnCl_2 + H_2 \qquad (9)$$
$$2NaOH + H_2SO_4 \rightarrow Na_2SO_4 + 2H_2O \qquad (10)$$
$$FeS + 2HCl \rightarrow FeCl_2 + H_2S \qquad (11)$$
$$2H_2S + O_2 \rightarrow 2H_2O + 2S \qquad (12)$$
$$Mg(OH)_2 + 2HNO_3 \rightarrow Mg(NO_3)_2 + 2H_2O \qquad (13)$$
$$Ca(OH)_2 + 2HCl \rightarrow CaCl_2 + 2H_2O \qquad (14)$$
$$2Na + 2H_2O \rightarrow 2NaOH + H_2 \qquad (15)$$
$$2Al(OH)_3 + 3H_2SO_4 \rightarrow Al_2(SO_4)_3 + 6H_2O \qquad (16)$$

In your revision of the reactions represented by the above equations you will have read about metals and acids in Chapter 7 for equation (9) and the iron and sulphur experiment in Chapter 5 for equations (11) and (12). Equations (10), (13), (14), and (16) are neutralization reactions (Chapter 7). Equation (15) represents the reaction of sodium with water (Chapter 14).

Mass calculations from equations

When we were discussing the equation for the concrete recipe (equation (5)) we found that the number of buckets of ingredients was equal to the number of buckets of concrete made. Obviously if we had weighed them, the total mass of ingredients and of concrete would also have been the same (before any water evaporated). One of the main uses of equations is to calculate the *relative* masses of substances taking part in a chemical reaction. On page 146 you will find a Table of Relative Atomic Masses of some of the common elements. You will learn more about atomic masses later in your chemistry, and all you need to know now is that they express the relative masses of atoms. Thus, the carbon atom (atomic mass 12) is twelve times heavier than the hydrogen atom (atomic mass 1), and the iron atom (atomic mass 56) is $1\frac{3}{4}$ times heavier than the sulphur atom (atomic mass 32). (How is the figure $1\frac{3}{4}$ obtained?) This, of course, means that a piece of iron (say a billion atoms of iron) weighs $1\frac{3}{4}$ times more than a piece of sulphur containing the same number of atoms; so if the piece of iron weighed 56 g the sulphur would weigh 32 g, or the iron might be 7 g and then the sulphur would weigh 4 g. By writing the atomic masses under the symbols and formulae of an equation, we can see the relative masses of substances taking part in the reaction. Let us take equation (2) for carbon and oxygen as an example:

$$C + O_2 \rightarrow CO_2$$
$$12 + 2 \times 16 \quad 12 + (2 \times 16)$$
$$\text{Totals} \quad 12 + \quad 32 \quad \quad 44$$

You will see that if we start with 12 g of carbon, 32 g of oxygen are needed to make 44 g of carbon dioxide. Or we can have a simpler ratio by dividing by 4:

$$C + O_2 \rightarrow CO_2$$
$$3\,g + 8\,g \quad 11\,g$$

Notice that there is no gain or loss in weight. The addition of the atomic masses in the molecules gives the *molecular masses*, 32 for oxygen and 44 for carbon dioxide. Other units of weight, e.g.

pounds or tons, could be used, but they are obsolescent now, and only the metric system is used in science.

We can refer to molecules and the molecular masses of carbon dioxide, water, and certain other compounds because they are made up of molecules. But there are many compounds which do not consist of molecules, and you will learn about their structure later. The formulae of these compounds indicate only the ratio of the different atoms in the compounds; for example, lead iodide (PbI_2) contains one lead atom to every two iodine atoms. In these types of compounds the addition of the atomic masses of the atoms in the formula gives the *formula mass* (the mass of one formula unit). So the formula mass of lead iodide is 461 (207 plus 2 × 127). At your present stage of chemistry you are not expected to know which compounds are composed of molecules, but you do know that electrolytes consist of charged atoms (ions, see Chapter 12), and therefore have formula masses and not molecular masses. The calculation of reacting masses from equations is just the same in each case.

Here is another example of these calculations to help you before you try some more unaided:

> What are the reacting masses when sulphuric acid neutralizes sodium hydroxide?

First, write the balanced equation:

$$H_2SO_4 + 2NaOH \rightarrow Na_2SO_4 + 2H_2O$$

Now write the atomic, molecular, or formula masses.

In this reaction we have compounds only, so we add up the separate atomic masses to give the formula masses (molecular mass for water):

$$H_2SO_4 + \quad 2NaOH \quad \rightarrow Na_2SO_4 + \quad 2H_2O$$
$$98 \qquad 2 \times 40 = 80 \qquad 142 \qquad 2 \times 18 = 36$$

Evidently 98 g sulphuric acid produce 142 g sodium sulphate. 80 g sodium hydroxide are needed to neutralize the acid, and 36 g water are also formed. From these figures you can work out any reacting mass by simple arithmetic. For example, how much sodium sulphate could you get from 4.9 g acid? And how much of the alkali would you need? How much acid would be needed to make 4.5 g water?

Now try the following questions:

1. What are the reacting masses in this reaction?:
$$Mg + 2HCl \rightarrow MgCl_2 + H_2$$

How much hydrogen could be obtained from 1.2 g of magnesium?

2. How much sulphur will be left over if 5.6 g iron are heated with 3.7 g sulphur to make iron sulphide? The equation is:

$$Fe + S \rightarrow FeS$$

3. If 50 g calcium carbonate ($CaCO_3$) are heated, how much quicklime (CaO) is obtained? How much slaked lime ($Ca(OH)_2$) could be produced from the quicklime? You should know the first equation. The second is:

$$CaO + H_2O \rightarrow Ca(OH)_2$$

You will now appreciate that a chemical equation is very different from a mathematical one. It describes details of an actual chemical reaction, the results of which can be verified experimentally (e.g. by careful weighing).

Calculating volumes of gases from equations

In the first of the above questions you found the mass of hydrogen produced, but it is much more useful to know the volume. This of course can be easily calculated if we know the density of the gas, but a quicker way is to remember that the gram molecular mass (molecular mass in grams) of any gas has a volume of 22.4 litres (at standard temperature and pressure). You will learn later the basis for this figure. In the example, 24 g magnesium produce 2 g hydrogen which is the gram molecular mass of hydrogen. This mass has a volume of 22.4 litres. From the above figures, 1.2 g magnesium ($\frac{1}{20}$ of 24 g) form 0.1 g ($\frac{1}{20}$ of 2 g) hydrogen or 1.12 litres ($\frac{1}{20}$ of 22.4 litres). (The answers to the other questions are 0.5 g sulphur, 28 g quicklime, and 37 g slaked lime.) You should now try the following questions, based on the fact that the gram molecular mass of any gas has a volume of 22.4 litres.

4. What volume of carbon dioxide can be obtained from 2.5 g of calcium carbonate?
(Reminder Write the equation and the formula/molecular masses under each compound.)

5. What mass of hydrogen peroxide (H_2O_2) must be decomposed to evolve 560 cm³ of oxygen? The equation is:

$$2H_2O_2 \rightarrow 2H_2O + O_2$$

Information given by equations

Now that you have become familiar with equations you should start writing the *state symbols* after each substance. This simply means that (s) is added after a solid, (l) after a liquid, (g) after a gas, and (aq), meaning aqueous, after a solution. Equations are not really complete without these additional symbols. Here are some examples

$$CaCO_3(s) \rightarrow CaO(s) + CO_2(g)$$
$$Zn(s) + 2HCl(aq) \rightarrow ZnCl_2(aq) + H_2(g)$$
$$NaOH(aq) + HCl(aq) \rightarrow NaCl(aq) + H_2O(l)$$

Note that acids are shown as solutions when they are mixtures of the pure compound and water. The amount of heat absorbed or evolved is also shown in equations when necessary, but you do not need to know how this is done at present. However, equations do not tell us everything we might want to know about a reaction, such as whether it is slow or vigorous, any colour changes, the necessary strength of an acid or alkali. So you should always describe a chemical reaction carefully and then write the equation on a line by itself.

Questions

1. A recipe for making dough for bread is to mix together 450 g flour (F), 7 g yeast (Y), 7 g sugar (S), 7 g salt (Sa) and 350 g water (W). Taking each symbol shown in brackets to represent one gram, write an equation for making dough (D).
2. Write the word equations for the following:
    (a) $S + O_2 \rightarrow SO_2$
    (b) $Ca + ZnO \rightarrow CaO + Zn$
    (c) $2H_2 + O_2 \rightarrow 2H_2O$
    (d) $CuO + H_2 \rightarrow Cu + H_2O$
    (e) $2Mg + O_2 \rightarrow 2MgO$
3. Balance the following skeleton equations and write the names of the substances under each symbol and formula:
    (a) $Cl_2 + NaI \rightarrow NaCl + I_2$
    (b) $HgO \rightarrow Hg + O_2$
    (c) $P_4 + O_2 \rightarrow P_4O_{10}$
    (d) $Cu(OH)_2 + H_2SO_4 \rightarrow CuSO_4 + H_2O$
    (e) $Al + HCl \rightarrow AlCl_3 + H_2$

Answers to questions 4 and 5, page 143: 4. 0.56 litre or 560 cm³, 5. 1.7 g.
Other answers have been given in the text.

4. Write balanced equations for the following reactions:
   (a) The decomposition of copper chloride into copper and chlorine.
   (b) The reaction between sulphur dioxide and water.
   (c) The neutralization of sodium hydroxide by sulphuric acid.
   (d) The reaction between iron and hydrochloride acid.
   (e) The burning of zinc.

5. Why must equations balance, and why cannot a formula be altered to help to balance an equation?

6. Do the total masses on one side of an equation always equal those on the other side? Describe an experiment by which you would try to prove this.

7. Write the reacting masses in the following equations (see page 146 for atomic masses):
   (a) $2Ca + O_2 \rightarrow 2CaO$
   (b) $CuCO_3 \rightarrow CuO + CO_2$
   (c) $Zn + PbO \rightarrow ZnO + Pb$
   (d) $Na_2CO_3 + 2HCl \rightarrow 2NaCl + H_2O + CO_2$
   (e) $Pb(NO_3)_2 + H_2SO_4 \rightarrow PbSO_4 + 2HNO_3$

8. (a) How much quicklime (calcium oxide) can be obtained by burning 2 g calcium?
   (b) What mass of sodium carbonate is needed to produce 11.2 litres carbon dioxide? (Use the equation in question 7 (d).)
   (c) How much sodium chloride could you make by neutralizing 1.6 g sodium hydroxide with hydrochloric acid?
   (d) How many cubic centimetres of hydrogen can be obtained if 1.3 g zinc react with sulphuric acid?
   (e) Gunpowder is a mixture of potassium nitrate (KNO_3), carbon, and sulphur. Its explosion can be represented by the following equation:

   $$2KNO_3 + S + 3C \rightarrow K_2S + 3CO_2 + N_2$$

   Calculate the total volume of gases produced by 33.75 g of gunpowder.

9. Write the word equations for the following reactions:
   (a) $3Fe(s) + 4H_2O(g) \rightarrow Fe_3O_4(s) + 4H_2(g)$
   (b) $CaCO_3(s) + 2HCl(aq) \rightarrow CaCl_2(aq) + H_2O(l) + CO_2(g)$

10. Write balanced equations, with state symbols, for the following reactions:
    (a) copper carbonate and dilute sulphuric acid; (b) the decomposition of mercury oxide (HgO).

Answers to question 8: (a) 2.8 g, (b) 53 g, (c) 2.34 g, (d) 448 cm³, (e) 11.2 litre.

Table of Formulae and Atomic masses

Formulae of Well-known Compounds

Name	Formula	Name	Formula
Water	H_2O	Copper chloride	$CuCl_2$
Aluminium oxide	Al_2O_3	Lead chloride	$PbCl_2$
Calcium oxide	CaO	Magnesium chloride	$MgCl_2$
Carbon dioxide	CO_2	Sodium chloride	$NaCl$
Carbon monoxide	CO	Copper nitrate	$Cu(NO_3)_2$
Copper oxide (black)	CuO	Magnesium nitrate	$Mg(NO_3)_2$
Magnesium oxide	MgO	Silver nitrate	$AgNO_3$
Mercury oxide	HgO	Sodium nitrate	$NaNO_3$
Sulphur dioxide	SO_2	Iron sulphide	FeS
Zinc oxide	ZnO	Calcium sulphate	$CaSO_4$
Calcium hydroxide	$Ca(OH)_2$	Copper sulphate	$CuSO_4$
Magnesium hydroxide	$Mg(OH)_2$	Magnesium sulphate	$MgSO_4$
Sodium hydroxide	$NaOH$	Sodium sulphate	Na_2SO_4
Calcium carbonate	$CaCO_3$	Lead iodide	PbI_2
Copper carbonate	$CuCO_3$	Hydrochloric acid	HCl
Sodium carbonate	Na_2CO_3	Nitric acid	HNO_3
Calcium chloride	$CaCl_2$	Sulphuric acid	H_2SO_4

Relative Atomic Masses of Common Elements

Aluminium (Al)	27	Manganese (Mn)	55
Bromine (Br)	80	Mercury (Hg)	201
Calcium (Ca)	40	Nitrogen (N)	14
Carbon (C)	12	Oxygen (O)	16
Chlorine (Cl)	35.5	Phosphorus (P)	31
Copper (Cu)	64	Potassium (K)	39
Hydrogen (H)	1	Silicon (Si)	28
Iodine (I)	127	Silver (Ag)	108
Iron (Fe)	56	Sodium (Na)	23
Lead (Pb)	207	Sulphur (S)	32
Magnesium (Mg)	24	Tin (Sn)	119
		Zinc (Zn)	65

Appendix

Safety Precautions

The object of the following paragraphs is to impress upon pupils the need for safety precautions in view of the potential danger in school laboratories. Teachers themselves are aware of the emphasis now necessarily being placed on safety by the Department of Education and Science, and will have studied the DES pamphlet 'Safety in School Laboratories'. Also recommended is 'Laboratory Safety—A Science Teacher's Source Book', by Armitage and Fasemore (Heinemann Educational Books, 1977).

Handling of Chemicals and Apparatus

Most chemicals are either *poisonous* or *injurious* in one way or another. You should therefore use only the smallest possible quantities in experiments, and exercise the greatest care in handling chemicals. Glass apparatus is a potential danger. If you use it carelessly and it breaks, the small sharp pieces of thin glass can cause deep cuts. And be very careful with Bunsen burners; their flames can be extremely hot. The following are some important **don't** rules for you to remember:

1. **Don't taste** any chemical unless you are instructed to do so by the teacher.
2. **Don't smell** any chemical unless similarly instructed. When smelling a gas or vapour never place the container directly under your nose, but use your hand to waft some of the gas from the container towards your nose. It is seldom necessary to smell chemicals, and those which have smells are often poisonous or injurious to the respiratory system.
3. **Don't forget to wash** your hands after touching chemicals and apparatus and before touching your face or food.
4. **Don't spill** chemicals or throw chemical waste into laboratory sinks. If you are in doubt or have an accident consult

your teacher. Solids are easily put into test tubes by means of a folded sheet of paper, and funnels can be used for liquids.

5. **Don't mix chemicals** together except as instructed by the teacher for a particular experiment. There can be dangerous results from mixing certain chemicals together. When taking some chemical from a reagent bottle or jar, *always replace the stopper at once*, otherwise the replacement of a stopper in the wrong container will cause contamination of the chemical concerned. Also, some chemicals should not be exposed to the air for longer than is necessary.

6. **Don't heat** chemicals more than is necessary. *Overheating can cause accidents.* Use a small Bunsen flame of moderate temperature. Be careful to ensure that the open end of a test tube is pointing in a safe direction in case your careless overheating causes a hot substance to shoot out of the test tube. It is usually advisable to use a test tube holder. The test tube (or other glass container) should always be moved gently in the flame in order to avoid excessive heat on one part of the glass, thereby preventing any cracking.

7. **Don't handle broken glass** with bare hands. Broken pieces of laboratory glassware are often thin and sharp, and the best way to gather them up is to use a dust pan and brush or improvised alternative.

8. **Don't forget the danger of fire,** and remember that a Bunsen flame is extremely hot (up to 1600 °C). Flammable substances should therefore be kept well away from a lighted Bunsen burner. It is important to adjust the Bunsen flame to give it a tip of yellow so that it is more easily seen. Otherwise it is easy to forget the presence of the hot flame and to catch your sleeve, etc., in it. This is more likely to happen in bright sunlight when the blue flame is very difficult to see.

9. **Don't forget the danger of hot apparatus, etc.** Do not touch hot tripod stands and wire gauzes, and be very careful when heating high boiling point substances, e.g. melted lead iodide for electrolysis.

10. **Don't fail to treat acids and alkalis with special care.** Alkalis can quickly injure the eyes.

11. **Don't cause accident risk** by running, pushing other pupils or getting in their way. *Never hurry.* Be careful and deliberate in everything you do.

12. **Don't fail to study carefully the rest of this guide.**

Acids and Alkalis

If strong, these compounds can be extremely dangerous. It is unlikely that you will be allowed to handle them, but this depends on your age and other factors. For most experiments only dilute solutions of acids and alkalis are needed, but even these can be corrosive towards skin and clothing and very injurious to eyes. Also, some acids evolve poisonous vapours when heated too much. You should always wash your hands after contact with acids and alkalis—indeed all chemicals—and sponge clothing with plenty of water to deal with acid or alkali spills.

If you should get any acid or alkali in your eyes or mouth, immediately splash them with large quantities of water and then go and tell your teacher. Never carry a bottle (large or small) of acid or alkali by the neck, but use both hands and be careful not to stumble over anything. If you are a senior pupil or laboratory assistant who is authorized to make a dilute acid from a concentrated one, always pour the **acid into the water. Never** add water to a concentrated acid because this can cause excessive heat which can generate a violent eruption of steam and acid.

Burns

Avoid burns by remembering not to touch hot apparatus, keeping your hands and arms clear of lighted Bunsen burners, and being extra careful when you are heating substances, especially hot liquids. Always hold a tripod stand by the ends of its legs after an experiment in case the stand is hot. If you receive a burn, go to the teacher at once for first aid.

Broken Glass

Many accidents are caused by broken glass. You should always be very careful when handling glass apparatus, particularly in the following circumstances.

(a) *'Sucking back'*

If a test tube or flask is used to decompose a chemical by heat and a delivery tube is fitted so that a gas can be collected over water, it is essential to *remove the delivery tube before stopping the heating*. If this is not done, the cooling gases in the tube or flask produce a partial vacuum (by their decrease in volume), resulting in cold water being 'sucked' into the hot glass container. At worst, this can have a very dangerous effect, depending on the chemical concerned. At best, the container is cracked and may break, its hot contents then spilling out.

(b) *Washing hot glassware*

Hot glass containers should not be washed with cold water, otherwise cracking may occur.

(c) *Fitting glass tubing to corks and bungs*

This should only be done under the supervision of your teacher. There are special safety precautions to be observed.

(d) *Use of pestle and mortar for breaking test tubes*

Always cover the test tube with a cloth.

(e) *Liquids boiling dry*

This can easily happen during distillation, evaporation, and when using a steam bath, and quickly causes the glass to crack. Always watch carefully the levels of liquids being heated.

Bibliography

1. *Secrets of Chemistry*, by Robert Brent (Hamlyn, 1965).
2. The following *Macdonald Junior Reference Library* books:
   Metals
   Gases
   Rocks and Minerals
   (Macdonald Educational)
3. The following *Nuffield Foundation Background Books*:
   The Chemical Elements—What they are and how they were discovered
   The Discovery of the Electric Current
   Burning
   Chemicals and where they come from
   Growing Crystals
   Petroleum
   Chemicals from Nature
   (Longman, 1966–8)

Index

(Bold figures indicate pages on which main treatment is to be found.)

152